A Colour Atlas of Fibreoptic Endoscopy of the Upper Respiratory Tract

A Colour Atlas of
Fibreoptic Endoscopy of the Upper Respiratory Tract

John D. Shaw

MB ChB (Sheffield), FRCS (England),
FRCS (Edinburgh)
Consultant in Otolaryngology,
Royal Hallamshire Hospital and The Children's Hospital,
Sheffield.
Honorary Clinical Lecturer, University of Sheffield.

Jack M. Lancer

MB ChB (Sheffield), LRCP (London), MRCS (England),
FRCS (England), DLO
Senior Registrar in Otolaryngology,
Royal Hallamshire Hospital and The Children's Hospital,
Sheffield.
Honorary Clinical Tutor, University of Sheffield.

Wolfe Medical Publications Ltd

Year Book Medical Publishers, Inc.

Copyright © J.D. Shaw, J.M. Lancer, 1987
Published by Wolfe Medical Publications Ltd, 1987
Printed by W.S. Cowell Ltd, 8 Butter Market, Ipswich, England
ISBN 0 7234 0887 4

For a full list of other atlases published by Wolfe Medical Publications Ltd,
please write to the publishers at: Wolfe House, 3 Conway Street, London, WIP 6HE,
England or Year Book Medical Publishers, Inc., 35 East Wacker Drive, Chicago,
Ill. 60601, USA

General Editor, Wolfe Medical Atlases: G. Barry Carruthers, MD (London)

Dr. ars Panda   mD, MRCP(UK).
12th Dec '88
London.

Distributed in Continental North America,
Hawaii and Puerto Rico by
Year Book Medical Publishers, Inc.

Library of Congress Cataloging in Publication Data

Shaw, John D.
    A colour atlas of fiberoptic endoscopy of the
upper respiratory tract.

    Includes index.
    1. Otolaryngologic examination–Atlases.
2. Otolaryngology–Diagnosis–Atlases.
3.Rhinolaryngoscopy–Atlases. 4. Fiber optics–
Atlases.   I. Lancer, Jack M.   II Title.
[DNLM: 1. Endoscopy–methods–atlases.   2. Fiber
Optics–instrumentation–atlases.   3. Respiratory
System–atlases.   4. Respiratory Tract Diseases–
diagnosis–atlases.   WF 17 S534c]
RF48.5.E53S49   1987   617′.5107′545   86-24702
ISBN 0-8151-7720-8

Contents

Acknowledgements

We thank Mr J.T. Buffin, Mr P.D. Bull, Mr R.T. Clegg and Mr D.F. Chapman, the consultant staff of the Sheffield Otolaryngology Department, who kindly allowed us to photograph patients under their care. Mr B.S. Crawford kindly advised us on the cleft palate illustrations. Thanks are also extended to the other medical, and the nursing and clerical staff of the department, all of whom contributed in some way to the production of this book.

Special thanks go to Sister Capper and her staff in the outpatient department, where the flexible fibreoptic rhinolaryngoscope is used; to Mr R. Turnstill and his staff in the Department of Medical Illustration at The Royal Hallamshire Hospital; and to the Key-Med company for their technical support and guidance.

Dedication

This book is dedicated to
Margaret and Lorraine

Introduction

Much of the work of the otolaryngologist involves the examination of the upper respiratory tract. Visualisation of the nasal and oral cavities is possible by direct examination in the outpatient department, whereas examination of the nasopharynx, larynx and laryngopharynx is usually possible only by indirect techniques.

Occasionally it may be impossible to examine these areas even after applying topical local anaesthetic because the patient is over-anxious, or has an easily excitable gag reflex, an overhanging epiglottis, or a narrow nasopharyngeal isthmus. If a thorough examination of these areas is considered necessary, conventional practice requires examination under anaesthesia. The flexible fibreoptic rhinolaryngoscope (FFRL) allows visualisation of the respiratory tract from the nasal cavity to the main bronchi under local anaesthesia in the outpatient department. If such an endoscopic examination shows no abnormality, there is then no need for hospital admission and general anaesthesia, thus saving a considerable amount of hospital time and expense. Since the introduction of the FFRL into this department, the number of examinations of the nasopharynx, larynx and laryngopharynx under general anaesthesia has halved.

The instrument used for examination and photography was the Olympus ENF-P (**A**).

A method of examination has been standardised in our department. After a full explanation of the procedure is given to the patient, the nasal cavity is anaesthetised with a 10% solution of topical cocaine, bilaterally if an examination of the nasal cavity is required, or the more patent side if the principal area to be examined is other than the nasal cavity.

A

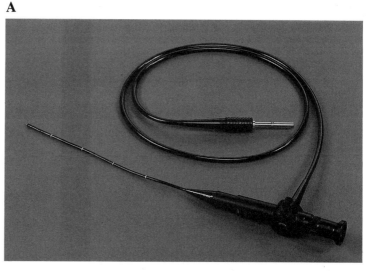

B

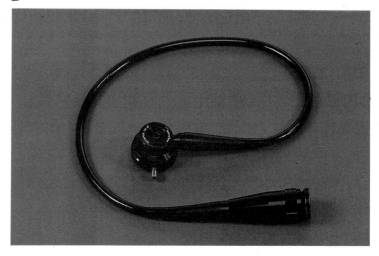

C

D

E

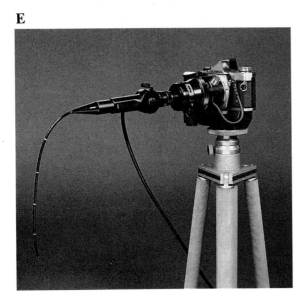

F

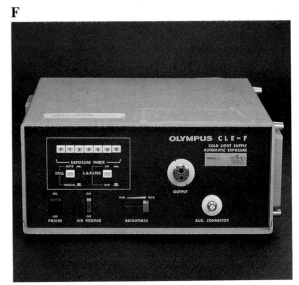

The patient is then given a benzocaine lozenge to suck, and if the subglottis and trachea are to be examined the vocal cords are sprayed with lignocaine solution. Ten minutes are allowed for the anaesthetic to take effect.

The patient lies supine on a couch with the head flexed to 30 degrees, and the examiner sits facing the patient. The instrument is introduced into the nasal cavity, and is then passed into the nasopharynx. By using a simple combination of movements of the patient's head from side to side, rotation of the shaft of the instrument between the thumb and forefinger, and employing the control on the body of the instrument which moves the tip, all areas are easily inspected. If appropriate, the patient may be allowed to see his own larynx, via a teaching attachment (**B**). This may be specially useful for functional disorders of the larynx.

The entire examination takes less than 2 minutes, and at the end of the examination the patient is advised to fast

for 2 hours to allow for recovery from the topical anaesthetic.

A method for photography was then devised. An Olympus OM-1n 35mm single lens reflex camera is used. The film used for this atlas was Kodak Ektachrome ED-135-36, which has an ASA rating of 200. The camera uses a clear glass focusing screen (1–9), which allows adequate light to reach the view-finder after passing through the fibrescope. The camera is attached to an automatic winder, (**C**), which allows the examiner's eye to remain focused on the subject and allows photography to proceed uninterrupted. A specialised adapter (Olympus SM-EFR 2) (**D**) connects the endoscope to the body of the camera. The adapter carries a focusing ring which should be set prior to undertaking any photography, and then should be maintained permanently at the same setting throughout all photography, unless the diopter setting on the endoscope has been altered – in which case a new focus setting will be required. The camera/winder combination is in turn mounted upon a tripod (**E**).

The eyepiece of the flexible fibreoptic rhinolaryngoscope is set to zero diopters, and if visual correction is required the examiner wears spectacles; alternatively, the diopter setting on the endoscope may be adjusted.

A synchronisation cord connects the adapter to the camera, and a camera cord connects the camera to the light source. The light source used is the Olympus CLE-F model, (**F**), which uses a 150 watt halogen lamp with mirror for general examination, and whose intensity may be varied. A xenon lamp provides the flashlight required for photography. The flash intensity is from a choice of 13 settings and should be varied during photography, depending on which area of the respiratory tract is being examined.

All photographs are taken with the camera mode setting on manual and with the shutter speed set to ¼ second. The apparatus is then ready for use and the examiner sits facing the patient (**G**).

Most of the photographs in this book have been taken in an outpatient clinic by the method described. However, a small number of photographs have been obtained by passing the FFRL through the oral cavity and directing its tip upwards towards the nasopharynx. This has allowed an improved view of the roof of the nasopharynx, and of the nasal cavity beyond the choanae from the nasopharynx, enabling the posterior ends of the turbinates to be examined.

In our experience, this endoscopic examination is difficult below the age of 8 years: thus conditions that occur exclusively in children below this age are not included.

G

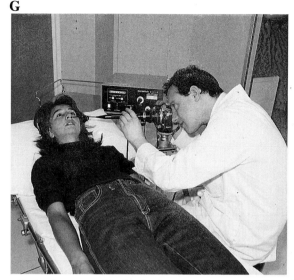

In the first part of the book, the normal anatomy of the upper respiratory tract is illustrated, while in its second part we have endeavoured to span the majority of disease processes that occur in this region. We realise that our account of disorders affecting the upper respiratory tract is not fully comprehensive, and we take responsibility for any omissions that may have occurred.

We hope that, as well as being of value to the otolaryngologist, the book will be valuable to the anaesthetist, the oral surgeon, the chest physician and surgeon, the plastic surgeon, the anatomist, the speech therapist, the medical student, and any other person who has an interest in endoscopy of the upper respiratory tract.

1 Normal Anatomy

Nasal Cavity

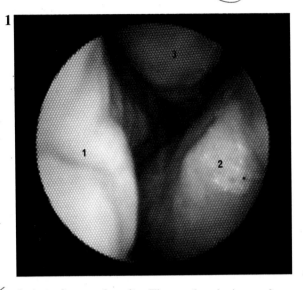

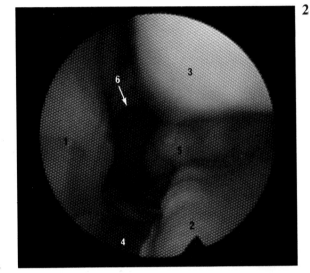

1 Anterior nasal cavity. The nasal cavity is seen from the nasal vestibule. The nasal septum (1) lies medially, and the inferior turbinate (2) lies laterally. The middle turbinate (3) is superior to the inferior turbinate.

2 Mid-nasal cavity. The nasal septum (1) is medial, and the inferior turbinate (2) and the middle turbinate (3) are lateral to the floor of the nose (4). The Eustachian (pharyngotympanic) tube (5) is visible beyond the choana (6).

3 Floor of nasal cavity. The nasal septum (1) is situated medially within the nasal cavity. Lateral to the floor of the nose (2) is the inferior meatus (3), which underlies the inferior turbinate (4). The lymphoid tissue of the nasopharynx lies beyond the choana (5). The Eustachian tube orifice (6) is distal to the posterior end of the inferior meatus.

4 Floor of nasal cavity. Sphenopalatine vessels course along the nasal floor.

5

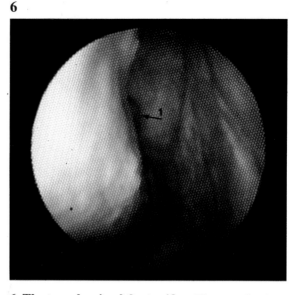

5 Posterior end of nasal cavity. The Eustachian tube orifice (1) is beyond the choana (2) in the nasopharynx. Prominent sphenopalatine vessels course along the nasal floor (3) and the Eustachian cushion (4). The nasal septum (5) lies medially, and the inferior turbinate (6) lies laterally within the nasal cavity.

6

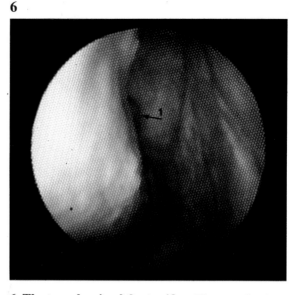

6 The naso-lacrimal duct orifice. The naso-lacrimal duct orifice (1) is high up and at the anterior end of the inferior meatus. This view shows it to be partially obscured by the inferior turbinate. The naso-lacrimal duct always drains into the inferior meatus, and its orifice may be slit-shaped or circular.

7

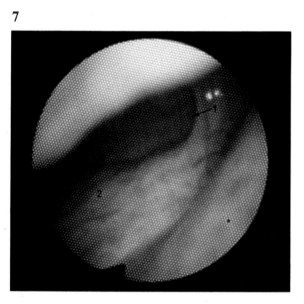

7 The naso-lacrimal duct orifice. This slit-shaped naso-lacrimal duct orifice (1) opens high within the inferior meatus (2).

8

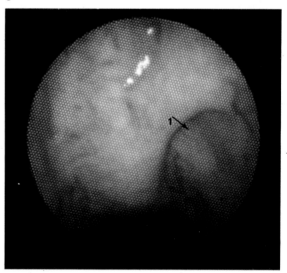

8 The naso-lacrimal duct orifice. This naso-lacrimal duct orifice (1) within the inferior meatus is slit-shaped.

9

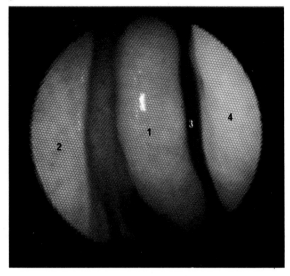

9 The middle turbinate. The middle turbinate (1) is viewed from below. The nasal septum (2) lies medially. Beneath the middle turbinate lies the middle meatus (3), which overlies the inferior turbinate (4).

10

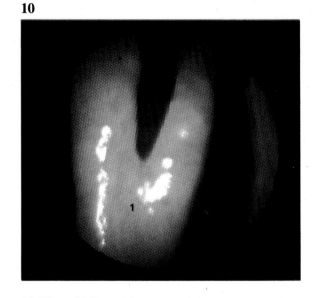

10 The middle turbinate – a variant. The root of the middle turbinate (1) bifurcates and surrounds an air cell ostium. The middle turbinate may occasionally contain a number of air cells.

11

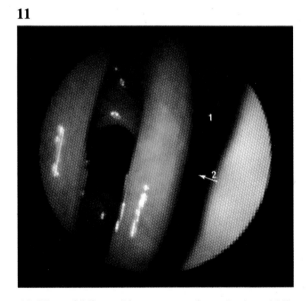

11 The middle turbinate – a variant. In the middle meatus (1) a further ethmoidal cell ostium (2) is present (see figure **10**).

12

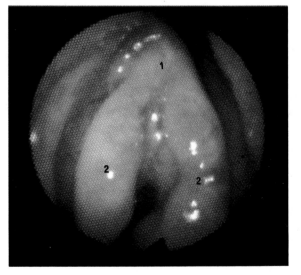

12 The middle turbinate – a variant. The middle turbinate bifurcates from a common stem (1) into two distinct limbs (2).

13

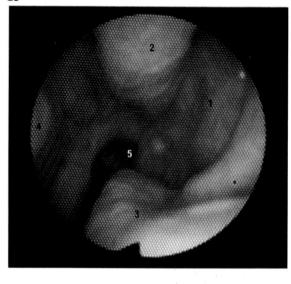

13 The middle meatus. The middle meatus (1) is the space between the middle turbinate (2) and the inferior turbinate (3), and lies laterally within the nasal cavity. The nasal septum (4) forms the medial border of the choana (5).

14

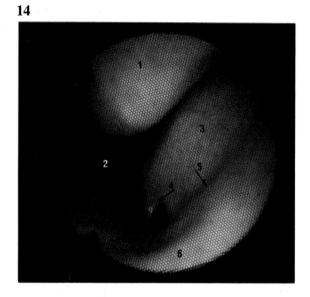

14 The middle meatus. Below the middle turbinate (1), in the middle meatus (2), is the prominence of the bulla ethmoidalis (3) containing middle ethmoidal air cells, draining via a common ostium (4) into the hiatus semilunaris (5). The inferior turbinate (6) forms the lower limit of the middle meatus.

15

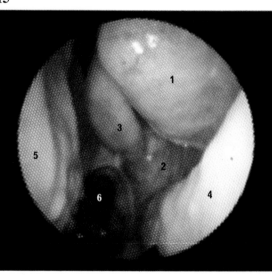

15 The bulla ethmoidalis. A prominent bulla ethmoidalis (1) within the middle meatus (2) lies beneath the middle turbinate (3), and above the inferior turbinate (4). Medially is the nasal septum (5), and distally is the choana (6).

16

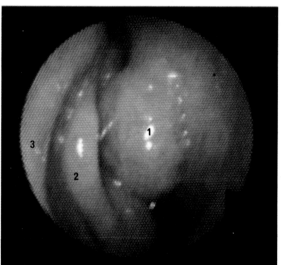

16 The bulla ethmoidalis. A prominent bulla ethmoidalis (1) fills most of the middle meatus, and indents the middle turbinate (2), which is in turn touching the nasal septum (3).

17

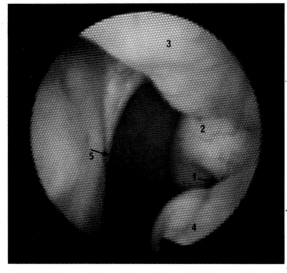

17 The Eustachian tube orifice. The Eustachian tube orifice (1) and cushion (2) are viewed from within the middle meatus, beneath the middle turbinate (3), and above the inferior turbinate (4). The posterior edge of the nasal septum (5) forms the medial border of the choana.

18

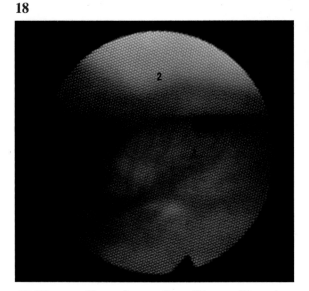

18 The maxillary sinus ostium. The maxillary sinus ostium (1) is partially obscured by the middle turbinate (2). This opening is at least partly through membrane (3). The appearance of the maxillary ostium may vary between individuals.

19

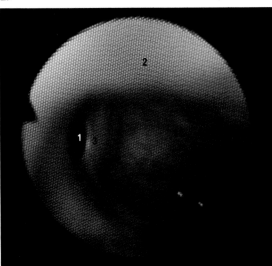

19 The maxillary sinus ostium. The opening of the maxillary sinus (1) is within the middle meatus and below the middle turbinate (2). This opening is at least partly through membrane (3).

20

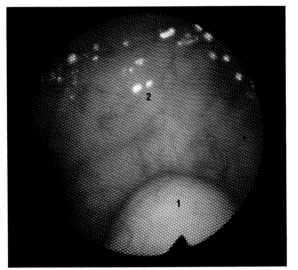

20 The maxillary sinus. The flexible fibreoptic rhino-laryngoscope (FFRL) has entered the maxillary sinus via its ostium; the sinus is lined by healthy mucous membrane throughout. The root of the second pre-molar tooth (1), which is covered by thin mucous membrane, is projecting into the maxillary antrum (2).

21

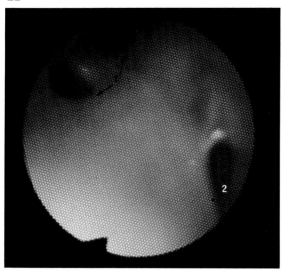

22

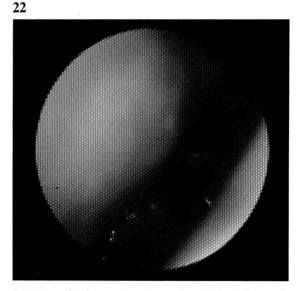

21 The middle meatus, with ethmoidal and maxillary sinus ostia. Two ethmoidal sinus ostia (1), and the maxillary sinus ostium (2) drain into the middle meatus.

22 The middle meatus. Two ethmoidal air cell ostia drain into the middle meatus.

23

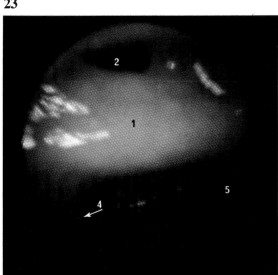

24

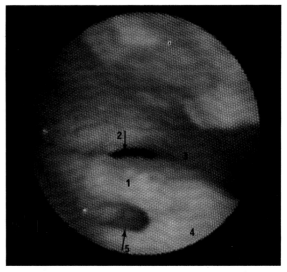

23 The common ostium of the posterior ethmoidal air cells. Above the middle turbinate (1), the posterior ethmoidal air cell ostium (2) drains into the superior meatus (3). The maxillary sinus ostium (4) lies within the middle meatus (5) (see figure **19**).

24 The superior turbinate. Above the superior turbinate (1), the sphenoidal sinus ostium (2) drains into the spheno-ethmoidal recess (3). Into the superior meatus (4), beneath the superior turbinate, a posterior ethmoidal air cell drains via its ostium (5).

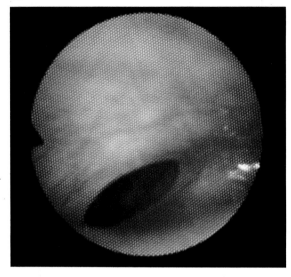

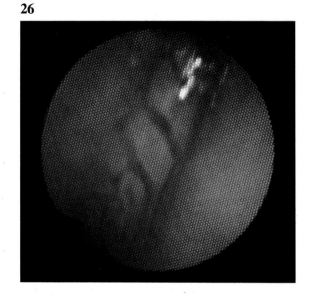

25 The sphenoidal sinus ostium. The sphenoid sinus drains via its ostium into the spheno-ethmoidal recess (see figure **24**).

26 The roof of the nose. This area, directly beneath the cribriform plate and through which olfactory nerve fibres leave the nose, represents the junction of the medial and lateral nasal walls. It is lined by a specialised neuro-epithelium.

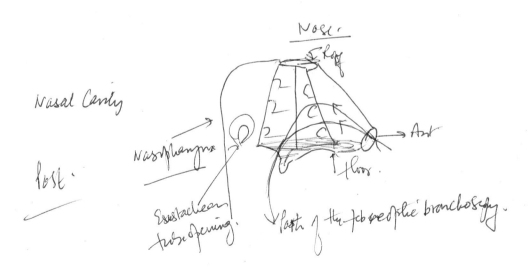

Nasopharynx

V-nrp view ✓

27

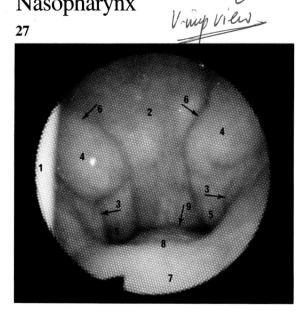

28

27 Nasopharynx. An overall view of the nasopharynx from the left choana shows the posterior edge of the nasal septum (1) medially.

On either side of the posterior wall of the nasopharynx (2) is a Eustachian tube orifice (3) partly surrounded by a prominent Eustachian cushion (4). Infero-medially lies the salpingo-pharyngeal fold (5) which is produced by the underlying salpingopharyneus muscle, and which arises from the posterior lip of the cartilage of the Eustachian cushion.

The fossa of Rosenmüller (pharyngeal recess) (6) is the space between the Eustachian cushion and the posterior wall of the nasopharynx.

The soft palate (7) and uvula (8) form the floor of the nasopharynx. The nasopharyngeal isthmus in this patient appears as a slit (9).

28 The choanae from within the nasopharynx. The posterior edge of the nasal septum (1) lies anteriorly and inferiorly to the roof of the nasopharynx (2), and forms the medial border of both the left (3) and right (4) choanae. Within both nasal cavities and beyond the choanae are the posterior ends of the inferior turbinates (5), and the middle turbinates (6).

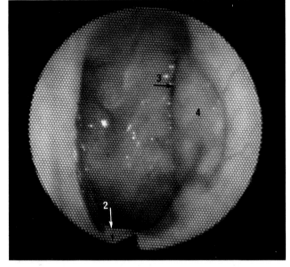

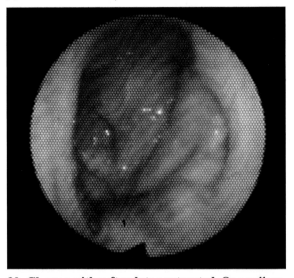

29 Choana with soft palate relaxed. The soft palate (1) is relaxed and therefore the nasopharyngeal isthmus (2) remains open. The fossa of Rosenmüller (3) is posterior to the Eustachian cushion (4).

30 Choana with soft palate contracted. On swallowing, the muscles of the soft palate (1) contract, and the nasopharyngeal ishthmus is closed. During this phase of swallowing, the Eustachian tube opens, allowing the pressure within the middle ear cavity to equalise to that within the nasopharynx (see figure **29**).

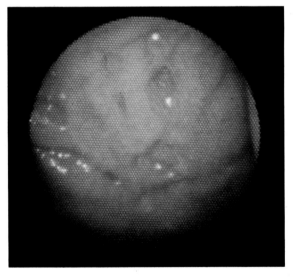

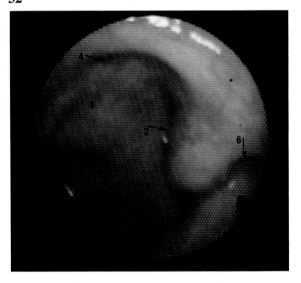

31 The nasopharyngeal tonsil. Several pits lie within regressing lymphoid tissue.

In the adult the nasopharynx has a smooth lining. Hypertrophic nasopharyngeal lymphoid tissue (adenoids) normally regresses around puberty, but may occasionally persist into adult life.

32 The lateral wall of the nasopharynx. The lateral wall of the nasopharynx (1) lies above the fossa of Rosenmüller (2) and continues toward the roof of the nasopharynx (3), which is supported by the basilar part of the occipital bone, and, to a lesser extent, by the posterior part of the body of the sphenoid.

The choanal edge (4) marks the boundary between the nasal cavity and the nasopharynx. The soft palate (5) leads towards the Eustachian tube orifice (6).

33

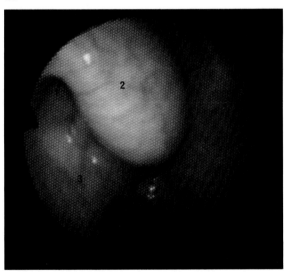

33 The fossa of Rosenmüller. The fossa of Rosenmüller (1) is the space between the Eustachian cushion (2) and the lateral wall of the nasopharynx (3).

34

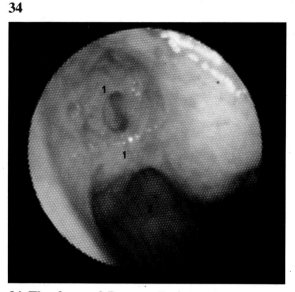

34 The fossa of Rosenmüller. Bridging webs (1) across the fossa of Rosenmüller (2) may be particularly prominent, and occur to varying degrees between individuals.

35

36

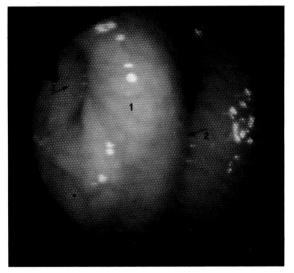

35 The Eustachian cushion. The Eustachian tube orifice (1) is surrounded by the Eustachian cushion (2) superiorly, and the soft palate (3) inferiorly.

The fossa of Rosenmüller (4) is posteromedial to the Eustachian cushion.

36 The Eustachian cushion. The Eustachian cushion (1) separates the fossa of Rosenmüller (2) medially, and the Eustachian tube orifice (3) laterally.

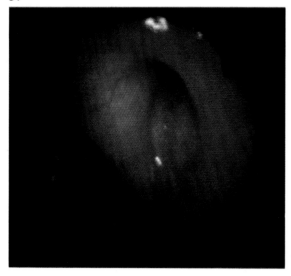

37 The Eustachian tube orifice and cushion. Mucus is draining from the Eustachian tube orifice (1).

There is a wide variation in the appearance of the Eustachian tube orifice and cushion between individuals.

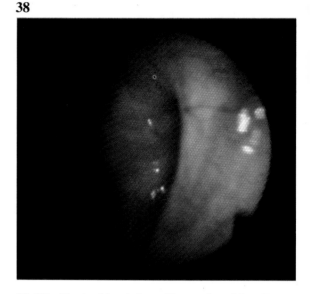

38 The Eustachian tube orifice and cushion. Prominent blood vessels overlie the Eustachian cushion.

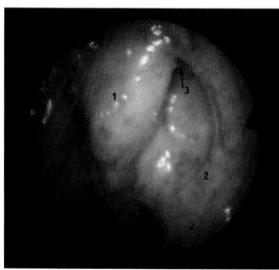

39 The Eustachian tube orifice and cushion. The Eustachian cushion (1) is particularly well developed in this patient who has a complete congenital palatal cleft. The soft palate muscles (2) lie anterior to the origin of the tube itself (3) (see figure **137**).

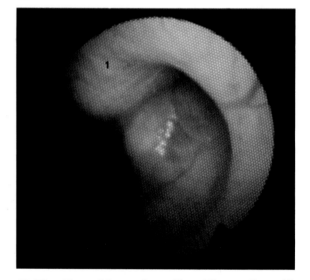

40 The Eustachian tube orifice. The cartilage (1) surrounding the Eustachian tube orifice is characteristically J-shaped. Part of the musculature of the soft palate originates from the lateral side of the cartilaginous Eustachian tube.

41

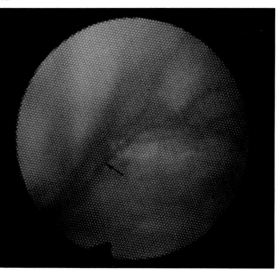

41 The closed Eustachian tube. When the muscles of the soft palate are relaxed, the Eustachian tube remains closed, and appears as a slit (arrow).

42

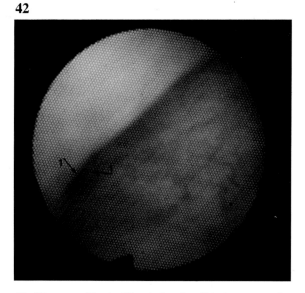

42 The open Eustachian tube. Swallowing causes the Eustachian tube to open (1). As part of the swallowing process, the muscles of the soft palate – levator and tensor palati – contract, and open the Eustachian tube. This enables the pressure within the nasopharynx to equalise to that within the middle ear cavity.

43

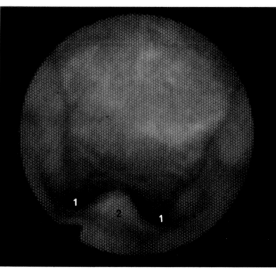

43 The nasopharynx. The nasopharyngeal isthmus (1) adopts a wide aperture because its muscles are relaxed. The uvula (2) is prominent.

Vimp view in FOB

44

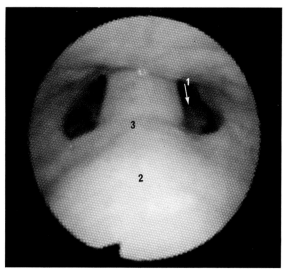

44 The relaxed soft palate. Beyond the naso-pharyngeal isthmus is the epiglottis (1). A prominent uvula leads from the soft palate (2), and a well-defined musculus uvulae (3) appears mid-way along the upper surface of the soft palate.

45

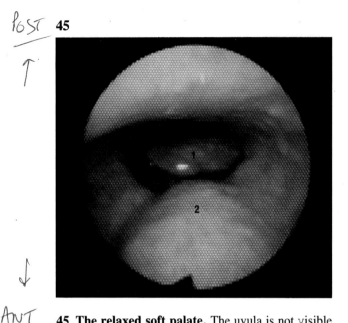

45 The relaxed soft palate. The uvula is not visible and there is an unobstructed view towards the epiglottis (1), beyond a relaxed soft palate (2).

46

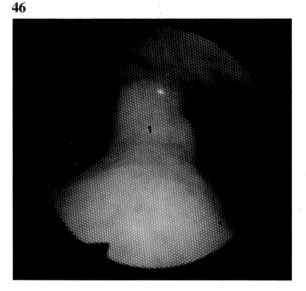

46 The uvula. A particularly elongated uvula (1) abuts against the posterior wall of the oropharynx (2).

47

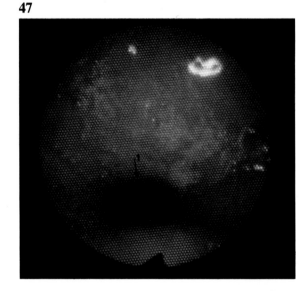

47 Passavant's ridge. Passavant's ridge (1) is produced by the contraction of those fibres of palatopharyngeus arising from the lateral aspect of the posterior part of the hard palate. It is against this ridge that the soft palate is elevated by the levator palati during phonation and swallowing.

The degree of prominence of the ridge varies between individuals.

48

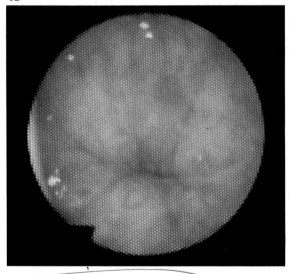

48 The nasopharyngeal sphincter. During swallowing and parts of speech the nasopharynx is closed off from the remainder of the respiratory tract by muscles of the soft palate and the pharynx. This prevents regurgitation of food and drink into the nasopharynx and nasal cavity, and also prevents escape of air during the production of certain sounds during speech.

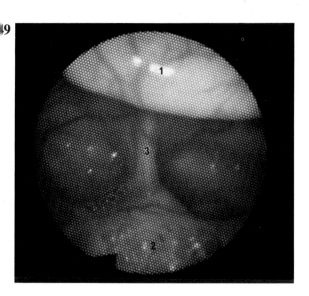

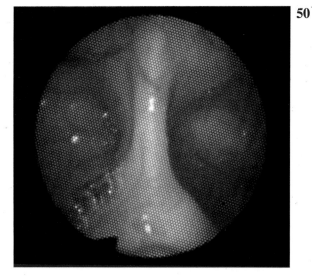

49 The vallecula. The vallecular space, the lowest limit of the oropharynx, lies between the laryngeal surface of the epiglottis (1) posteriorly, and the base of the tongue (2) anteriorly. This space is bisected by the median glosso-epiglottic fold (3).

50 The median glosso-epiglottic fold. This midline fold of mucous membrane extends between the base of the tongue and the epiglottis.

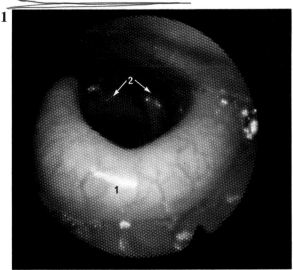

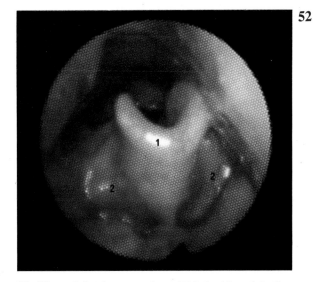

51 The epiglottis – a variant. This incidental variation of an epiglottis (1) has a thickened and rolled margin. Beyond the epiglottis are the arytenoid cartilages (2).

52 The epiglottis – a variant. This incidental finding of an infantile epiglottis (1) in an adult was asymptomatic. The epiglottis is more acutely folded than in the more common adult epiglottis, and tends to fall posteriorly, making laryngeal examination by indirect laryngoscopy difficult. Thus, in patients with suspected laryngeal pathology, examination with the FFRL becomes an essential outpatient procedure. This narrow epiglottis has created a wide vallecular space (2).

53

POST

opening into the Esophagus. here.

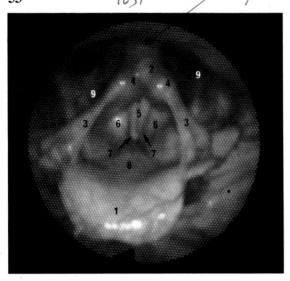

ANT

53 An overall view of the larynx during phonation. The inlet of the larynx is bounded by the epiglottis (1) anteriorly and the arytenoid cartilages (2) posteriorly. These structures are connected by the ary-epiglottic folds (3), which contain the cuneiform cartilages (4) within their upper borders.

Above the adducted vocal cords (5) are the ventricular bands (false vocal cords, vestibular folds) (6). The space between the vocal cords and the ventricular bands is the laryngeal sinus (laryngeal ventricle) (7). The epiglottic tubercle (8) is at the base of the epiglottis, above the anterior commissure. The pyriform fossae (9) lie posterior to the ary-epiglottic folds.

54

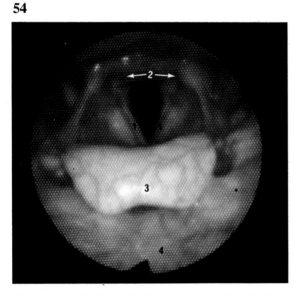

55

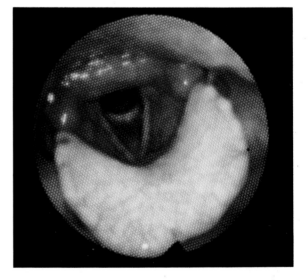

54 An overall view of the larynx during inspiration. During inspiration the vocal cords (1) are abducted, and the inter-arytenoid space (2) is opened. The epiglottis (3) curls towards the base of the tongue (4). This is the commonest appearance of the epiglottis.

55 An overall view of the larynx during inspiration. This epiglottis has a U-shaped configuration and curls in a horizontal plane towards the base of the tongue at its midpart, but curls in a vertical plane away from the tongue at its margins.

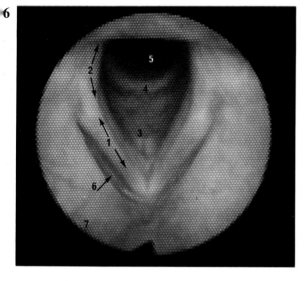

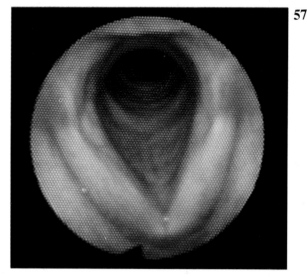

56 The rima glottidis. The rima glottidis is the space between the free edges of the vocal cords (1) and the vocal processes of the arytenoid cartilages (2).

Below the vocal cords are the crico-thyroid membrane (3), the cricoid cartilage (4), and the tracheal rings (5) encircling the tracheal lumen.

The laryngeal sinus (6) lies between the vocal cords and the ventricular bands (7).

57 The vocal cords. The vocal cords are the free upper borders of the conus elasticus, and are formed by the sharp reflection of the mucous membrane over the vocal ligaments.

(Rico-pharyngeus. opeing of post. Esophagushen

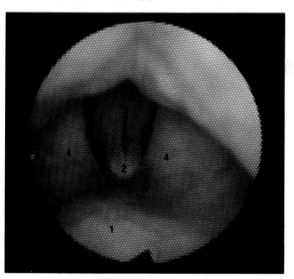

Insp. view

Ant

58 The vocal cords during phonation. The epiglottic tubercle (1) lies above the anterior commissure (2).

During phonation the rima glottidis is reduced to a slit (3) and the ventricular bands (4) have no active role, and remain abducted.

This is not the case in the pathological condition of dysphonia plicae ventricularis, when the ventricular bands are the predominant structures in voice production (see figure **218**).

59 The arytenoid cartilages during phonation. During phonation the inter-arytenoid muscle contracts, approximating the arytenoid cartilages (1) and obliterating the inter-arytenoid space (2).

Each cuneiform cartilage (3) lies towards the posterior end of the ary-epiglottic fold (4).

Lateral to each ary-epiglottic fold is the pyriform fossa (5), and the postero-lateral wall of the laryngopharynx (hypopharynx) (6).

60

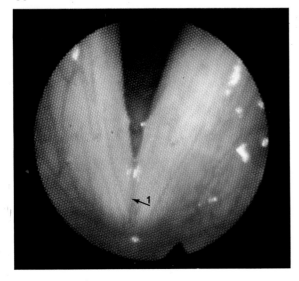

60 The anterior commissure. The anterior commissure (1) marks the junction of the anterior ends of both vocal cords.

61

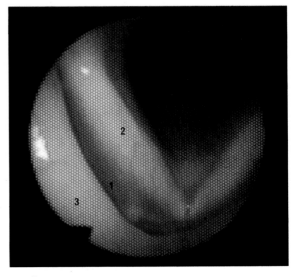

61 The right vocal cord and anterior commissure. The laryngeal sinus (1) lies above the vocal cord (2), and extends laterally beneath the ventricular band (3).

62

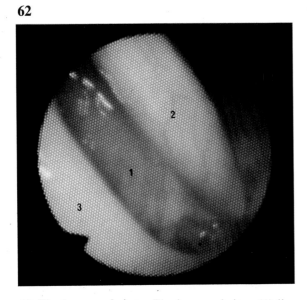

62 The laryngeal sinus. The laryngeal sinus (1) lies between the vocal cord (2) and the ventricular band (3).

63

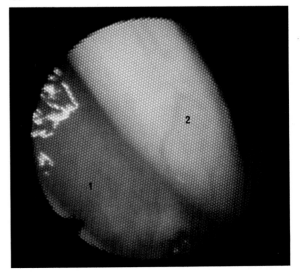

63 The laryngeal sinus. A healthy and delicate mucous membrane lines the laryngeal sinus (1) and is continuous with that of the vocal cord (2).

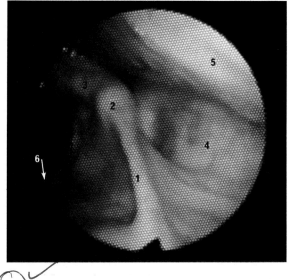

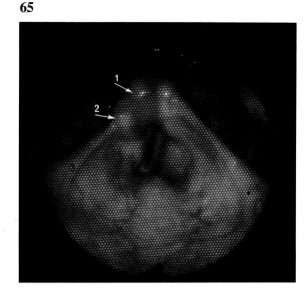

64 The left ary-epiglottic fold. The ary-epiglottic fold (1) is the free upper edge of the quadrate membrane, and contains muscle fibres which connect the side of the epiglottis to the muscular process and posterior surface of the opposite arytenoid cartilage.

Contraction of this muscle during swallowing is a part of the protective sphincteric action of the laryngeal inlet.

The cuneiform cartilage (2) and the corniculate cartilage (3) are prominent.

The pyriform fossa (4) and the lateral pharyngeal wall (5) are lateral to the ary-epiglottic fold. The anterior commissure (6) and the left vocal cord (7) lie medial to the ary-epiglottic fold.

65 The accessory laryngeal cartilages. The corniculate cartilage of Santorini (1) articulates with the summit of the arytenoid cartilage, while the club-shaped cuneiform cartilage of Wrisberg (2) lies within the upper border of the ary-epiglottic fold (3).

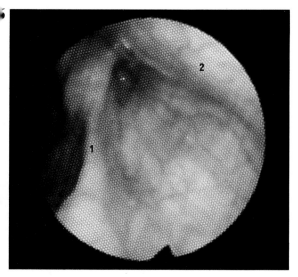

66 The left pyriform fossa. The pyriform fossa is a part of the laryngopharynx and is bordered medially by the quadrate membrane and ary-epiglottic fold (1), and laterally by the mucous membrane overlying the thyroid cartilage (2).

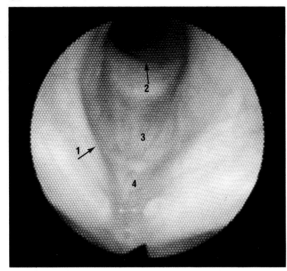

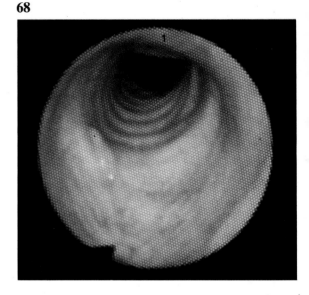

67 The subglottic space. The subglottic space is circumferential and lies between the under-surface of the vocal cords (1) and the lower border of the cricoid cartilage (2).

The mucous membrane of the subglottis covers the cricoid cartilage, the crico-thyroid membrane (3), and the lower part of the thyroid cartilage (4).

68 The trachea. Tracheal rings surround the lumen of the trachea except posteriorly (1) where the wall is completed by the trachealis muscle.

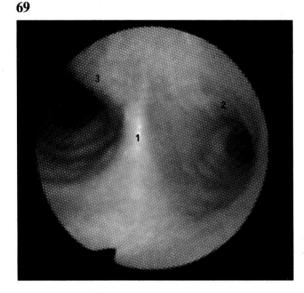

69 The carina. This view of the carina (1) and the bifurcation of the trachea into left (2) and right (3) main bronchi has been obtained via a tracheostomy. The right main bronchus lies in a more vertical plane than the left main bronchus. For this reason, inhaled foreign bodies lodge more frequently in the right side of the bronchial tree.

2 Disease Processes

Nasal Cavity

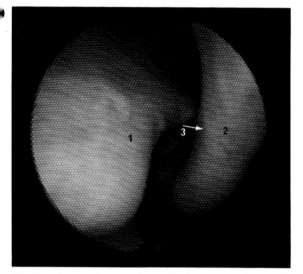

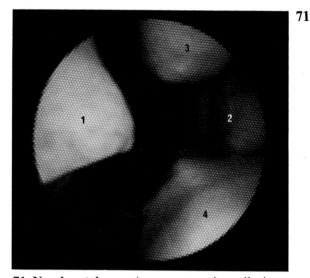

70 Nasal septal spur. A cartilaginous septal spur (1) partially occludes the left nasal cavity.

The inferior turbinate (2) has developed an indentation (3) in order to accommodate the septal deformity. Obstructive symptoms resulting from such a deformity may be corrected by the operation of submucosal resection of the nasal septum (SMR), or septoplasty.

71 Nasal septal spur. An asymptomatic cartilaginous septal spur (1) has its apex at the level of the middle meatus (2). The middle turbinate (3) lies superiorly and the inferior turbinate (4) lies inferiorly to the middle meatus.

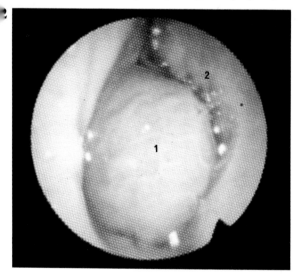

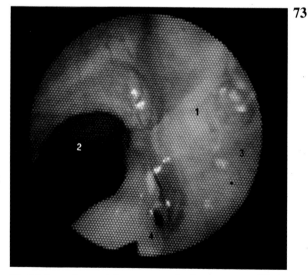

72 Acute rhinitis. Anterior rhinoscopy reveals a hyperaemic and oedematous inferior turbinate (1) with a surrounding clear mucoid discharge, due to coryza.

The inferior turbinate is abutting against the nasal septum (2).

73 Rhinorrhoea. A stream of clear mucus (1) is passing into the nasopharynx (2) along the middle meatus (3), above the inferior turbinate (4). The normal clearance of mucus from the nasal cavity and sinuses is by ciliary action beating towards the nasopharynx.

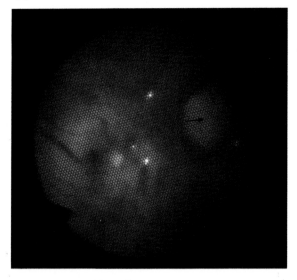

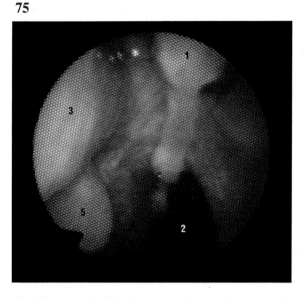

74 Acute sinusitis. The FFRL has passed through the maxillary sinus ostium into the maxillary antrum, which is lined by a hyperaemic mucous membrane, as a result of a recent upper respiratory tract infection. A small amount of mucopus is present (arrow).

75 Chronic rhinitis. A stream of pus (1) flows into the nasopharynx (2) above the middle turbinate (3) and Eustachian tube (4). Beneath the middle turbinate is a benign nasal polyp (5).

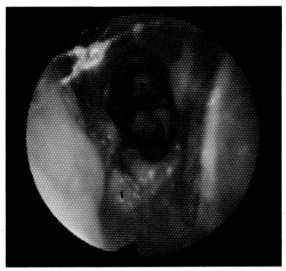

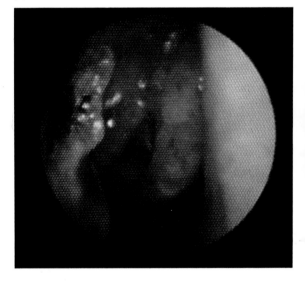

76 Atrophic rhinitis. Atrophic rhinitis is characterised by a thin, dry and inflamed mucous membrane. There are dry crusts along the nasal floor (1) and in the nasopharynx (2), and a mucus web bridges the nasal cavity (3). In this uncommon condition, a constant feature is a voluminous nasal cavity due to atrophy of the mucous membrane throughout. Atrophic rhinitis is usually idiopathic, but may also occur following extensive nasal surgery, or radiotherapy.

77 Atrophic rhinitis. Dry crusts overlie the turbinates (see figure **76**).

78

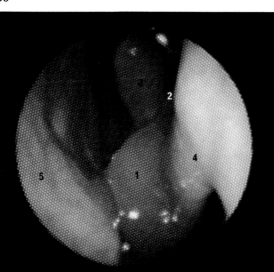

79

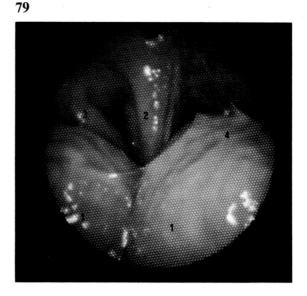

78 Allergic rhinitis. Anterior rhinoscopy reveals a hypertrophied inferior turbinate (1), with a characteristic blue appearance. This swollen turbinate is abutting against the nasal septum (2).

79. Allergic rhinitis. Posterior rhinoscopy reveals huge, blue hypertrophied inferior turbinates (1) which are protruding into the nasopharynx. The inferior turbinates lie on either side of the posterior end of the nasal septum (2). A normal middle turbinate (3) lies within the left nasal cavity. Copious clear mucus (4) is a characteristic feature.

80

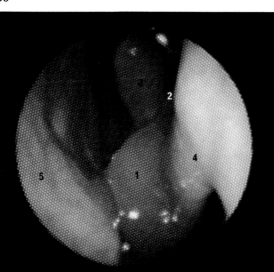

81

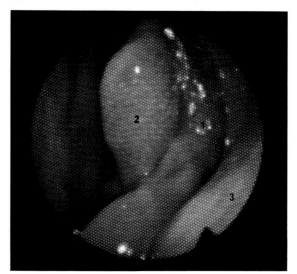

80 A nasal polyp. A nasal polyp (1) is arising from the anterior ethmoidal group of air cells and passes into the middle meatus (2), beneath the middle turbinate (3), and above the inferior turbinate (4). The polyp is abutting against the nasal septum (5). Nasal polyps consist of oedematous tissue arising more commonly from the ethmoidal air cells, and may occur as a response to allergy or infection.

81 The stalk of a nasal polyp. The stalk of the nasal polyp (1) lies beneath the middle turbinate (2), and above the inferior turbinate (3).

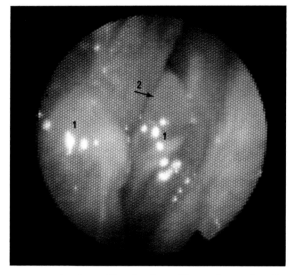

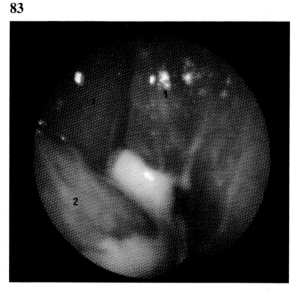

82 Nasal polyps. Nasal polyps (1) are obstructing the left and right nasal cavities, and lie on either side of a large cartilaginous nasal septal perforation (2).

83 Nasal polyps causing sinusitis. Two nasal polyps (1) are obstructing sinus ostia in the middle meatus. Sinusitis has resulted, with pus flowing over the inferior turbinate (2).

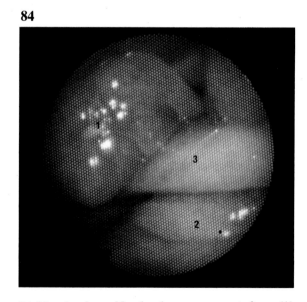

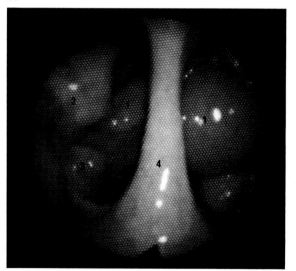

84 Nasal polyps. Nasal polyps are present above (1) and below (2) the middle turbinate (3). The polyp in the superior meatus is arising from the posterior ethmoidal group of air cells, and the polyp in the middle meatus is arising from the anterior or middle group of ethmoidal air cells.

85 Nasal polyps. Nasal polyps (1) fill both nasal cavities and are arising from the middle meatus between the middle turbinate (2) and the inferior turbinate (3). The polyps are passing towards the nasopharynx and are approaching the posterior end of the nasal septum (4), formed by the vomerine bone.

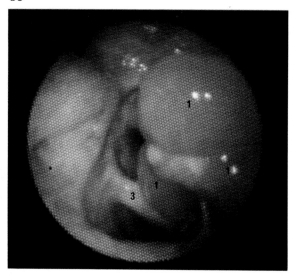

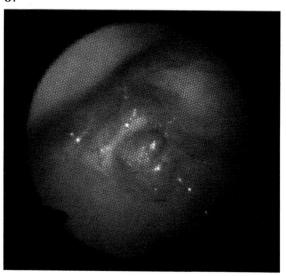

86 Nasal polyps in the nasopharynx. Nasal polyps (1) have passed beyond the choanae into the nasopharynx. The inferior border of the posterior edge of the nasal septum (2) is visible. The polyps in the left nasal cavity lie within the middle meatus, above the inferior turbinate (3).

87 Open ethmoidal air cells. This is the appearance within the middle meatus following nasal polypectomy, in which the ethmoidal air cells have been partially exenterated.

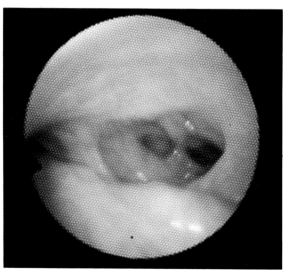

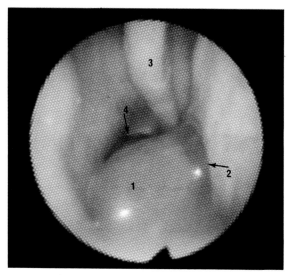

88 Open ethmoidal air cells. Following nasal polypectomy, there are enlarged openings into individual ethmoidal air cells within the middle meatus.

89 Antro-choanal polyp. A polyp (1) protrudes through its natural ostium (2) into the middle meatus, beneath the middle turbinate (3), and passes backwards towards the choana (4), almost filling the nasal cavity. The maxillary sinus ostium is enlarged owing to the pressure exerted by the protruding polyp.

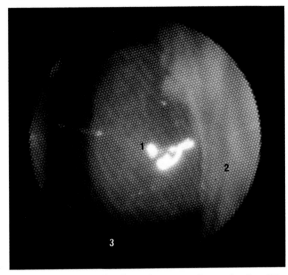

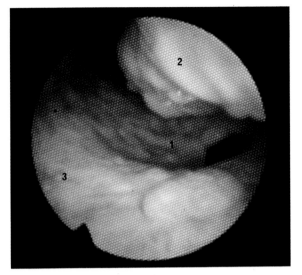

90 Antro-choanal polyp. An antro-choanal polyp (1), arising in the left maxillary antrum, is viewed from the right nasal cavity. The polyp is obstructing the right choana, whose boundaries include the nasal septum (2) and the soft palate (3). The fundus of an antro-choanal polyp usually passes through the choana and may fill the whole of the nasopharynx and the opposite choana, and may protrude below the soft palate into the oropharynx when it is visible through the mouth (see figure **89**).

91 Intra-nasal antrostomy. A smooth perforation lies within the inferior meatus (1), below the lower border of the inferior turbinate (2), and above the floor of the nose (3). This iatrogenic fistula allows communication between the nasal cavity and the maxillary sinus, and is usually performed for the treatment of maxillary sinusitis. This intra-nasal antrostomy has a small diameter as it was performed many years previously, and has undergone subsequent contraction.

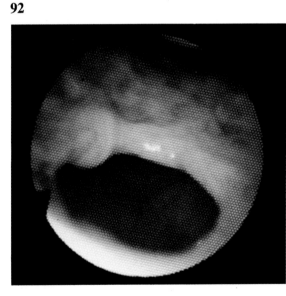

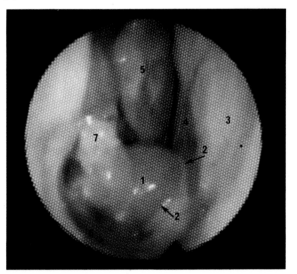

92 Intra-nasal antrostomy. Closer inspection from the nasal cavity reveals a healthy mucous membrane (1) within the maxillary antrum (see figure **91**).

93 Antral polyp. A benign nasal polyp (1) passes into the nasal cavity through an intra-nasal antrostomy (2) in the inferior meatus (3), below the inferior turbinate (4), and the middle turbinate (5). The polyp is lying upon the nasal floor (6) and is covered by mucus (7).

94

95

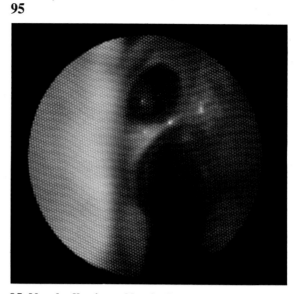

94 A nasal adhesion. A nasal adhesion (1) straddles the nasal cavity between the nasal septum (2) and the inferior turbinate (3). The middle turbinate (4) is above the adhesion.

95 Nasal adhesions. Nasal adhesions may be a complication of surgery or radiotherapy.

96

97

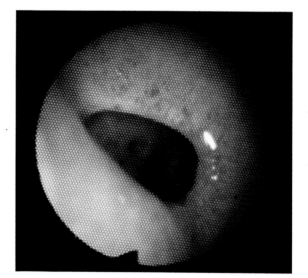

96 Nasal septal perforation. This small septal perforation (1) has a clean margin, and is situated towards the posterior end of the nasal septum (2), lying at the level of the middle turbinate (3).

The commonest septal perforations seen in otolaryngological practice follow nasal septal surgery.

97 Nasal septal perforation. The right nasal cavity is seen from the left nasal cavity through a clean nasal septal perforation.

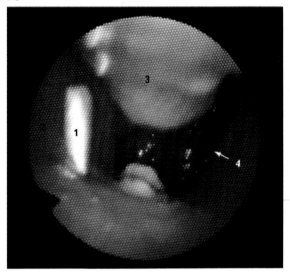

98 Perforation of the hard palate. The floor of the nasal cavity is the hard palate. This patient has a perforation of the hard palate; for the purpose of demonstration, a metal probe (1) has been passed into the nasal cavity, from the oral cavity, through the perforation. The probe lies within the inferior meatus (2), beneath the inferior turbinate (3). Along the floor of the nose lies food debris, which has entered the nose through the perforation during a meal. The posterior edge of the nasal septum (4) forms the medial border of the choana. This rare condition may occur with neoplasms and chronic inflammatory disorders, or may follow surgery to this area. The main problems encountered with such patients are difficulties with speech, and regurgitation of food and drink into the nasal cavity.

99

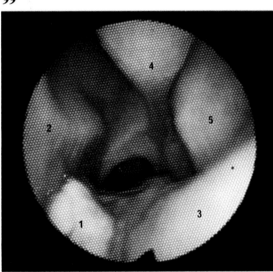

100

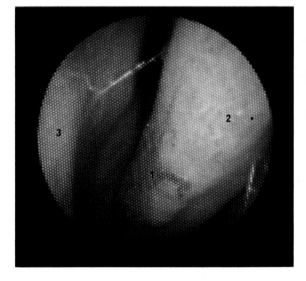

99 Rhinolith. A rhinolith (1) lies along the nasal floor, between the nasal septum (2), and the inferior turbinate (3), but beneath the middle turbinate (4) and the middle meatus (5).

Rhinoliths are irregular calcific masses which have usually been present for a prolonged period, and which usually occur around a nidus of foreign material.

100 Foreign material on the inferior turbinate. A grinder has inhaled metallic dust (1) which has become adherent to the mucus over the inferior turbinate (2). The nasal septum (3) is unaffected.

101

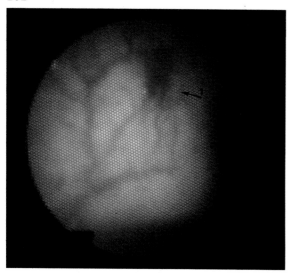

101 Abnormal nasal septal blood vessels. Abnormal blood vessels, lying over the anterior part of the cartilaginous nasal septum are converging to a point (1) which has burst and produced an epistaxis. This is the commonest site of bleeding in the nose, and is known as Little's area. At this point there occurs an anastomosis of the major blood vessels supplying the nasal septum (Kiesselbach's plexus).

102

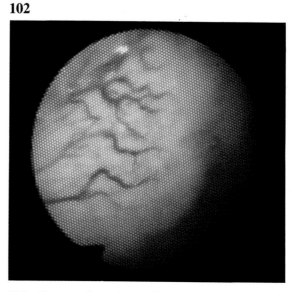

102 Abnormal nasal septal blood vessels. Following cocainisation of the nasal septum, the site of epistaxis is easily identified.

103

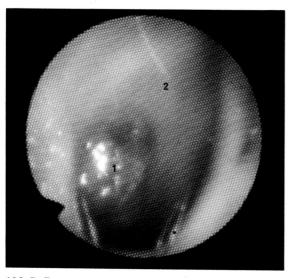

103 Inflammatory granuloma of the nasal septum. A benign granuloma (1) at the anterior end of the nasal septum (2) has caused a severe epistaxis.

Following its removal under local anaesthesia, the patient had no further symptoms.

104

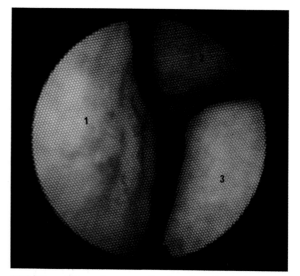

104 Hereditary haemorrhagic telangiectasia of the nasal septum. This systemic disease may cause epistaxes owing to the rupture of telangiectasia, here situated within the nasal septum (1), which is medial to the middle turbinate (2) and the inferior turbinate (3) (see figures **105, 127, 159, 195**).

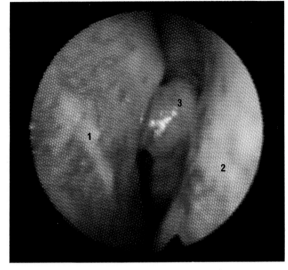

105 Hereditary haemorrhagic telangiectasia of the nasal cavity. Telangiectasia are within the nasal septum (1), the inferior turbinate (2), and the middle turbinate (3), and were the cause of recurrent epistaxes in this patient (see figures 104, 127, 159, 195).

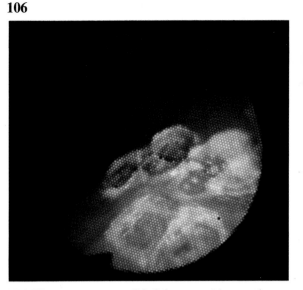

106 Electrocautery to Little's area. Abnormal vessels which produced an epistaxis have been treated by electrocautery under local anaesthesia (see figure 102).

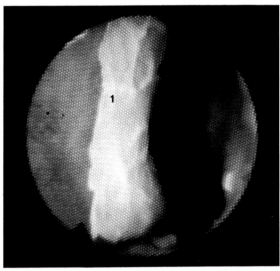

107 Chemical cautery to Little's area. Silver nitrate may be used to cauterise chemically blood vessels that are the cause of an epistaxis.

This treated area on the nasal septum (1) is pale, while the adjacent area on the nasal septum is healthy.

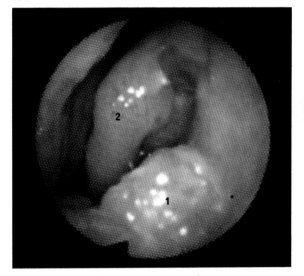

108. Inverted nasal papilloma (Ringertz tumour). The tumour (1) is in the middle meatus, beneath the middle turbinate (2).

Although benign, this fleshy looking growth tends to recur locally and therefore a wide excision is recommended.

109

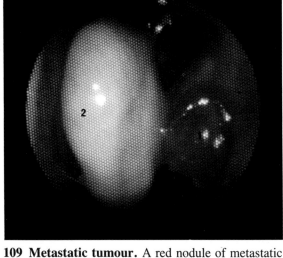

109 Metastatic tumour. A red nodule of metastatic squamous cell carcinoma (1) lies in the middle meatus beneath the middle turbinate (2). The primary site was the nasopharynx (see figure **148**).

110

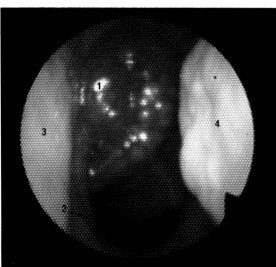

110 Metastatic tumour. A nodular metastatic squamous cell carcinoma from the nasopharynx (1) is passing through the choana (2) into the nasal cavity, and impinges on the nasal septum (3) and the inferior turbinate (4).

111

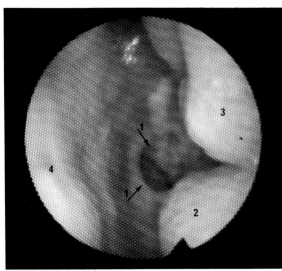

111 Choanal stenosis. A reduced choanal aperture (1) lies beyond the inferior turbinate (2), the middle turbinate (3), and the nasal septum (4).

112

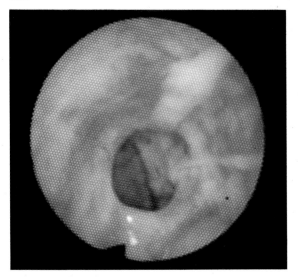

112 Choanal stenosis. The stenotic aperture is rounded, and its margins are covered by a pale and atrophic mucous membrane. This elderly patient had received radiotherapy to the nasopharyngeal area many years previously. This appearance is more commonly seen as a congenital abnormality, or following surgery for congenital choanal atresia (see figure **111**).

Nasopharynx

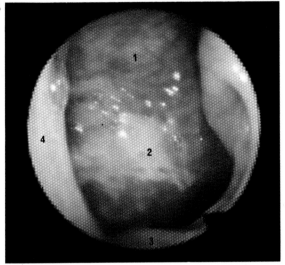

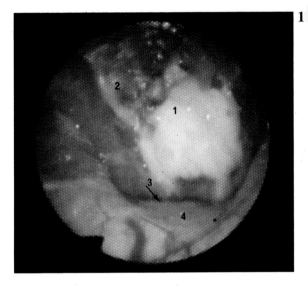

113 Chronic nasopharyngitis. The mucous membrane of the nasopharynx (1), beyond the choana, is chronically inflamed, and its central part is covered by drying mucopus (2).

The boundaries of the choana include the soft palate (3), and the nasal septum (4).

114 Chronic nasopharyngitis. Mucopus (1) and crusts (2) overlie the nasopharyngeal mucous membrane, and are encroaching towards the nasopharyngeal isthmus (3), posterior to the soft palate (4).

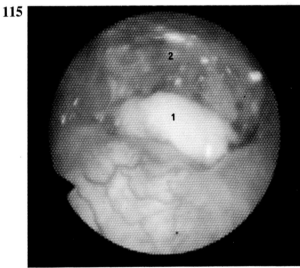

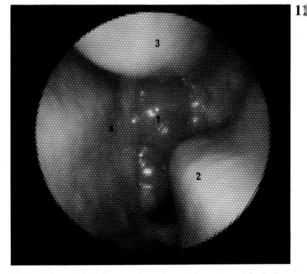

11

115 Chronic nasopharyngitis. Mucopus (1) and crusts (2) are occluding the nasopharyngeal isthmus.

116 Adenoidal hypertrophy. Hypertrophied lymphoid tissue (nasopharyngeal tonsil, adenoids) (1) is almost obstructing the choana, and abuts against the inferior turbinate (2), the middle turbinate (3), and the posterior edge of the nasal septum (4).

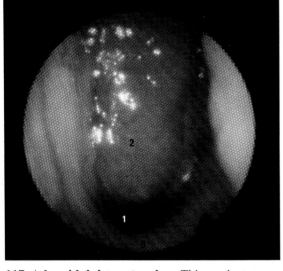

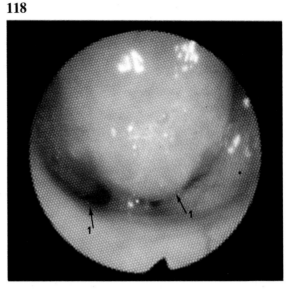

117 Adenoidal hypertrophy. This patient complained of nasal obstruction because of a reduced nasal airway (1) due to enlarged adenoidal tissue. The reduced airway lies between the lower border of the adenoids (2) and the soft palate (3).

118 Adenoidal hypertrophy. A narrow slit (1) may be all that remains to allow for nasal ventilation and drainage. As a consequence, the patient is predisposed to rhino-sinusitis, and may complain of snoring, mouth breathing, and nasal discharge.

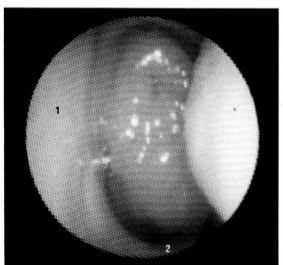

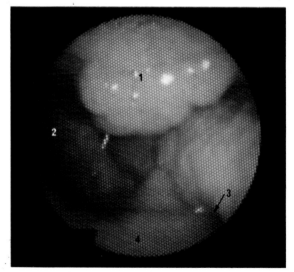

119 Adenoidal hypertrophy. The enlarged adenoid adapts to the contours of the choana, which include the nasal septum (1) and the soft palate (2).

120 Regressing adenoidal tissue. A small adenoid pad (1) is all that remains in this adolescent and a good airway is present within the nasopharynx (2). Laterally lies the Eustachian tube orifice (3), and anteriorly is the soft palate (4).

It is usual for adenoidal tissue to regress around puberty.

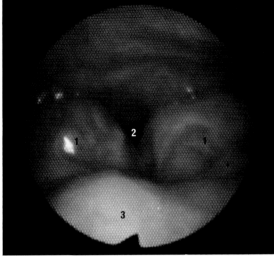

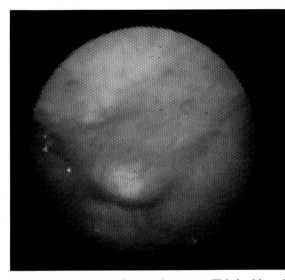

12

121 Tonsillar hypertrophy. Gross hypertrophy of the palatine tonsils (1) has resulted in partial obstruction of the nasopharyngeal airway (2). The tonsils lie below the soft palate (3), and are impinging upon the posterior wall of the nasopharynx. This appearance may not be evident on oral examination.

Upper airway obstruction due to adenotonsillar hypertrophy may result in sleep apnoea, especially in children.

Symptomatic obstruction to this degree is an indication for tonsillectomy, even in the absence of inflammation.

122 Nasopharyngeal retention cyst. This incidental finding was asymptomatic.

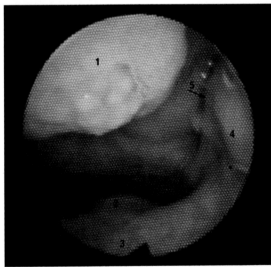

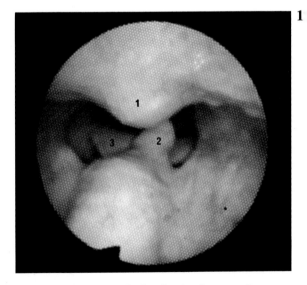

123 Tornwaldt cyst. This cyst (1) has originated from the junction of the roof and posterior wall of the nasopharynx, and is an embryological remnant of Rathke's pouch.

The cyst lies above the uvula (2) and the soft palate (3), and medially to the Eustachian tube orifice (4) and the fossa of Rosenmüller (5).

124 Prominent cervical spine in the nasopharynx. This asymptomatic swelling (1) in an elderly man is the anterior tubercle of the arch of the atlas, and is posterior and superior to the uvula (2). The epiglottis (3) lies distally.

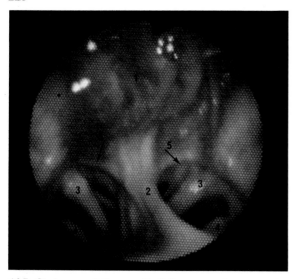

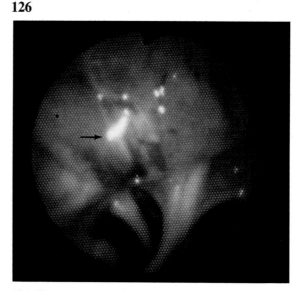

125 Oedema of the roof of the nasopharynx. An air-gun pellet has penetrated the roof of the nasopharynx, and has caused an acute inflammatory reaction (1). Below this area is the posterior end of the nasal septum (2). Healthy middle turbinates (3) and inferior turbinates (4) lie within the nasal cavity beyond the choanae (5).

126 Foreign body of the roof of the nasopharynx. Several weeks later, the air-gun pellet had begun to extrude spontaneously (arrow) (see figure **125**).

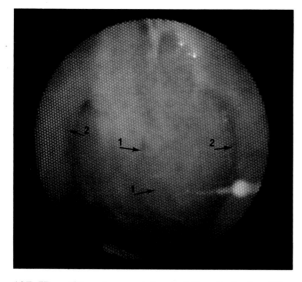

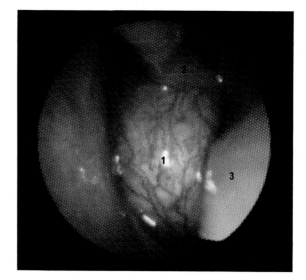

127 Hereditary haemorrhagic telangiectasia of the nasopharynx. Telangiectasia (1) lie within the mucous membrane of the nasopharynx, and are medial to the fossa of Rosenmüller (2). This systemic disease may affect any area of the respiratory tract (see figures **104, 105, 159, 195**).

128 Juvenile angiofibroma. An expanding lesion (1) has moulded itself to the contours of the middle turbinate (2), and the inferior turbinate (3). This large benign tumour has its origin in the nasopharynx, and slowly enlarges over many years. It produces symptoms of nasal obstruction and bleeding.

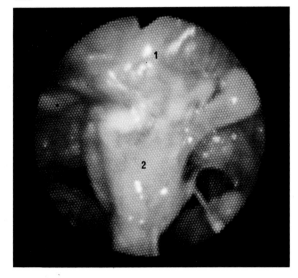

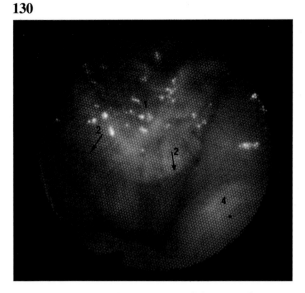

129 Squamous cell carcinoma of the nasopharynx.
A squamous cell carcinoma of the roof of the nasopharynx (1) is spreading towards the nasal cavity, and encroaches along the posterior border of the nasal septum (2).

130 Squamous cell carcinoma of the nasopharynx.
The tumour (1) has involved most of the posterior wall of the nasopharynx. Its advancing margin (2) has a raised and rolled edge. The soft palate (3) appears unaffected, but the left Eustachian cushion (4) is oedematous. As a consequence, Eustachian tube dysfunction will ensue with the formation of a middle ear effusion causing deafness. This may be the presenting feature of the disease (see figure **129**).

Eustachian tube

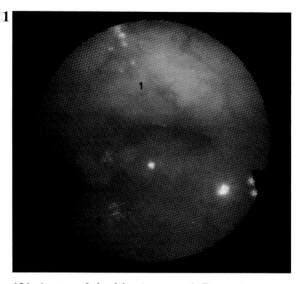

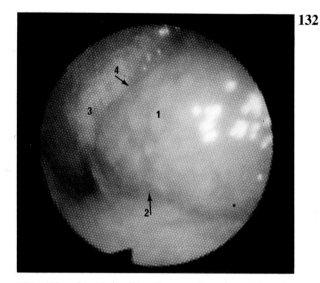

131 Acute salpingitis. An acute inflammatory reaction has resulted from recent radiotherapy to the nasopharynx. The inflammation extends from the Eustachian cushion (1) to the soft palate (2).

132 Allergic salpingitis. Gross oedema overlying the Eustachian cushion (1) is occluding the Eustachian tube orifice (2), and clear mucus (3) is flowing along the fossa of Rosenmüller (4) and the nasopharynx. Secretory otitis media may occur in any disease affecting the Eustachian tube orifice. A conductive deafness would then usually ensue.

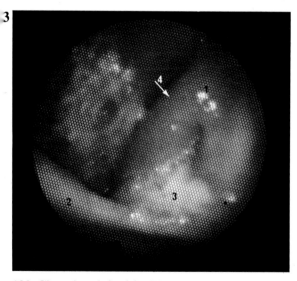

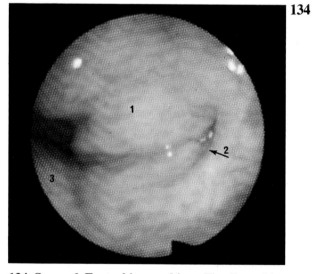

133 Chronic salpingitis. The Eustachian cushion (1) is inflamed and oedematous owing to disordered nasopharyngeal ventilation and drainage, as a result of a palatal fenestration (2) for a malignant growth. Mucopus (3) issues from the Eustachian tube orifice, and a dry crust lies on the posterior wall of the nasopharynx and extends into the fossa of Rosenmüller (4).

134 Scarred Eustachian cushion. The Eustachian cushion (1) has been removed by over-zealous adenoidectomy several years previously. The Eustachian tube orifice (2) is slit-shaped and extends towards the soft palate (3).

135

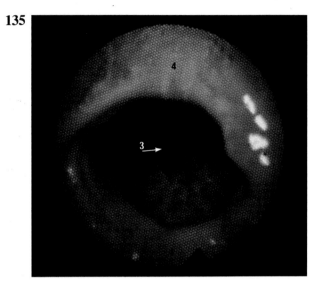

135 Congenital cleft palate. This large congenital palatal defect is in an elderly patient who had no corrective surgery as a child. The cleft is incomplete and the patient has an intact hard palate (1). Beyond the defect is the dorsum of the tongue (2), and epiglottis (3). Clear mucus (4) streams down the posterior wall of the nasopharynx.

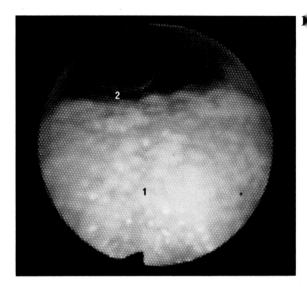

136 Dorsum of tongue. This unusually good view of the dorsum of the tongue (1), from the nasal cavity, can only be obtained through a palatal defect. The tip of the epiglottis (2) is in contact with the base of the tongue (see figure **135**).

137

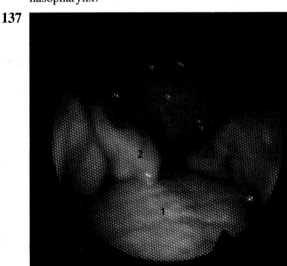

137 Complete congenital palatal cleft. This elderly patient has had a complete palatal cleft throughout her life. The dorsum of the tongue (1) lies below the free edges of the palatal remnants (2), and anterior to the posterior wall of the nasopharynx (3) (see figure **39**).

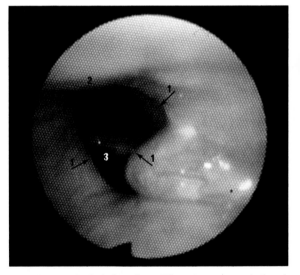

138 Congenital cleft palate. There is a congenital and asymmetrical cleft of the soft palate, the margins of which consist of soft palate muscles on three sides (1) and the posterior pharyngeal wall on the fourth side (2). Beyond the cleft is the epiglottis (3). This appearance has resulted from a breakdown of the surgical repair.

139

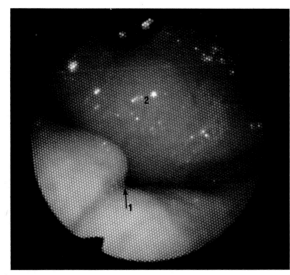

140

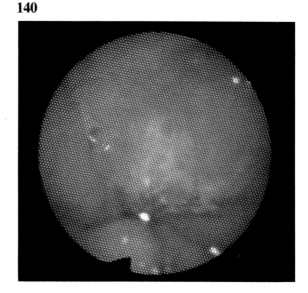

139 Repaired cleft palate. The posterior limit of the palatal repair (1) remains as a notch. A shallow groove marks the line of the repair (2).

140 Repaired cleft palate during swallowing. The same patient is asked to swallow and there is complete closure of the nasopharyngeal isthmus (see figure **139**).

141

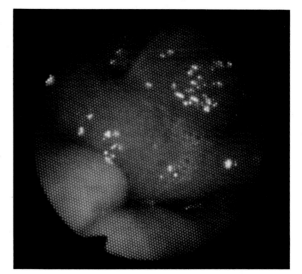

142

141 Repaired cleft palate. The posterior limit of the repair remains as a notch (1), and a large adenoid pad (2) overlies the posterior wall of the nasopharynx.

142 Repaired cleft palate during swallowing. The same patient is asked to swallow. The adenoid pad contributes to the closure of the nasopharyngeal isthmus by coming into contact with the soft palate muscles during their contraction. For this reason, the adenoids should be preserved in patients with a congenital cleft palate (see figure **141**).

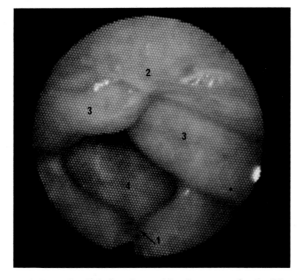

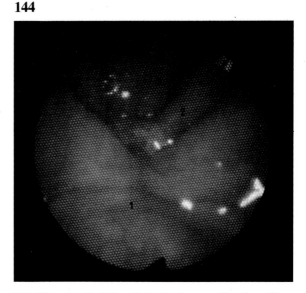

143 Repaired cleft palate with pharyngoplasty. The posterior limit of the repaired cleft palate appears as a dimple (1), and forms the anterior limit of the nasopharyngeal isthmus. The posterior wall of the nasopharynx (2) has been augmented by two flaps of muco-muscular tissue from the palatopharyngeal muscles (3) (Hynes' pharyngoplasty). Beyond the nasopharyngeal isthmus is the tongue (4).

144 Repaired cleft palate during swallowing. During swallowing, the nasopharynx is completely closed off from the oropharynx, following repair of a congenital cleft palate. The nasopharyngeal isthmus consists of the soft palate anteriorly (1), the muscular bar of the pharyngoplasty (2), and an adenoid pad posteriorly (3).

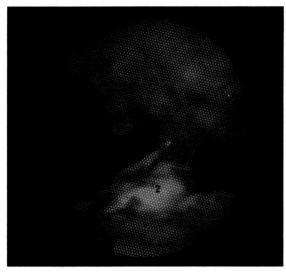

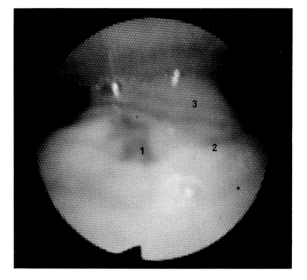

145 Burns on the uvula. Following accidental ingestion of a caustic corrosive, an acute inflammatory reaction of the uvula (1) has developed. The uvula, which is covered with an exudate (2), is impinging upon the posterior pharyngeal wall (see figures **170, 207**).

146 Telangiectasis of the uvula. A solitary telegiectasis (1) is at the junction of the soft palate (2) and the uvula (3).

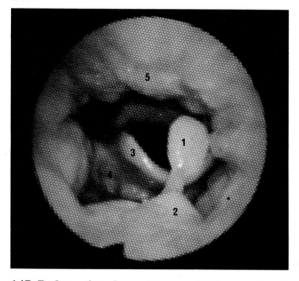

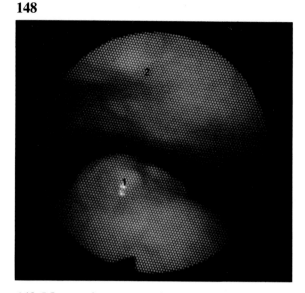

147 Pedunculated papilloma of the uvula. A pedunculated papilloma (1) hangs free from the tip of the uvula (2). Beyond the papilloma is the epiglottis (3), vallecula (4), and the posterior wall of the oropharynx (5).

148 Metastatic tumour of the uvula. The nodular lesion (1) on the dorsum of the uvula is a metastatic squamous cell carcinoma. The site of origin of this tumour is the nasopharynx. The posterior wall of the nasopharynx (2) at this level is healthy (see figure **109**).

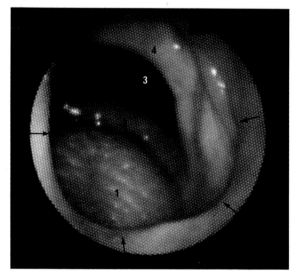

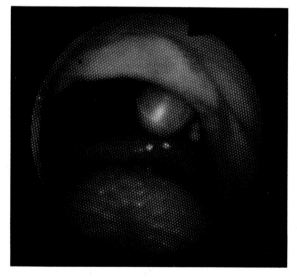

149 The oral cavity. This patient has undergone a partial maxillectomy, resulting in a large palatal fenestration, the margins of which are arrowed. The dorsum of the tongue (1) lies below the lower lip (2), which forms the lower border of the mouth (3), and whose upper border is the upper lip (4).

150 The mouth. For the purpose of demonstration, the patient has placed a finger onto the lower lip (see figure **149**).

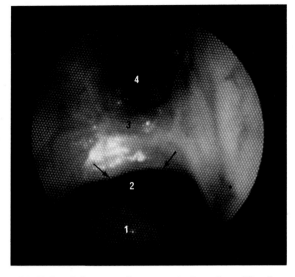

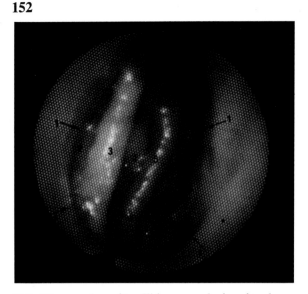

151 Palatal fenestration – posterior view. The dorsum of the tongue (1), and epiglottis (2), are viewed from within the nasal cavity, through a large palatal fenestration, the posterior margin of which is arrowed. The soft palate (3) is between the fenestration and the choana (4) (see figure **149**).

152 The nasal cavity. This unusual view has been obtained by passing the FFRL through the mouth into the oral cavity, and directing its tip upwards towards the nasal cavity. Beyond the margins of the palatal fenestration (1), are the inferior meatus (2), the inferior turbinate (3), the middle meatus (4), the middle turbinate (5), and the nasal septum (6) (see figure **149**).

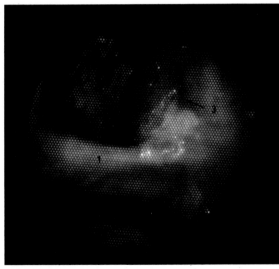

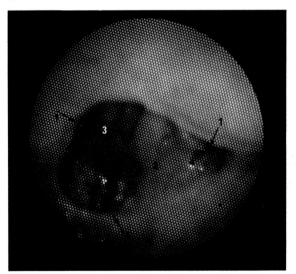

153 Palatal fenestration – posterior view. Above the posterior margin of the palatal fenestration (1), is the Eustachian cushion (2), and the Eustachian tube orifice (3), which is draining mucus. Dry crusts lie on the posterior wall of the nasopharynx (4). The dorsum of the tongue (5) is visible in the oral cavity (see figure **133**).

154 Palatal fenestration. Beyond the margin of the palatal fenestration (1) is the lateral wall of the nose (2), which separates the nasal cavity (3) medially from the maxillary antrum (4) laterally (see figure **153**).

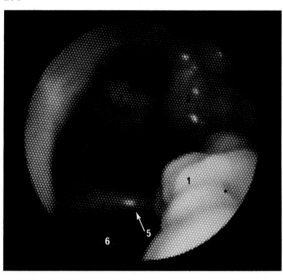

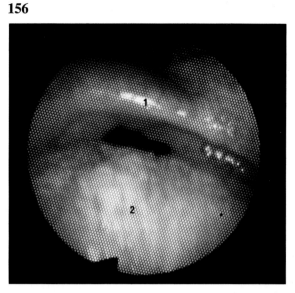

155 Palatal fenestration. A 'mulberry' hypertrophy of the inferior turbinate (1) has probably occurred as a result of disordered ventilation of the nasal cavity, subsequent to an acquired palatal fenestration. The Eustachian cushion (2) and the nasopharynx (3) lie above the soft palate (4), which forms the posterior margin of the fenestration (5). The dorsum of the tongue (6) lies below the fenestration (see figure **153**).

156 Palatal fenestration. Below the posterior margin of the palatal fenestration (1) is the dorsum of the tongue (2) (see figure **153**).

Tongue and epiglottis

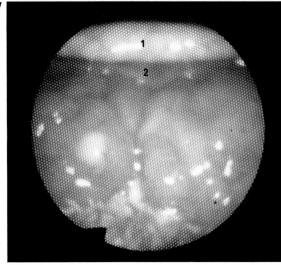

157 Hairy tongue. Part of the posterior one-third of the tongue anterior to the epiglottis (1), and above the vallecula (2), is coated and hairy.

This occurs more commonly following radiotherapy to this area, in patients taking prolonged courses of antibiotics, or in heavy smokers.

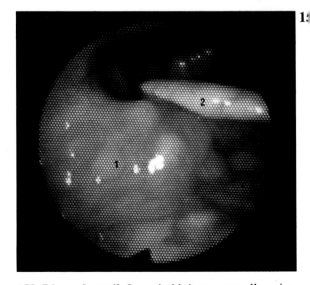

158 Lingual tonsil. Lymphoid tissue normally exists on the base of the tongue as the lingual tonsil, and may occasionally hypertrophy. Because it is prominent in this patient (1), it is abutting against the tip of the epiglottis (2). Occasionally lingual tonsillitis may occur during an upper respiratory tract infection and cause similar symptoms to palatine tonsillitis.

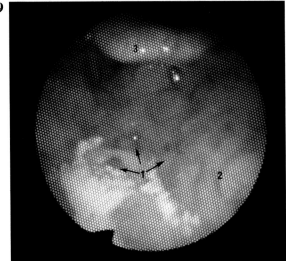

159 Hereditary haemorrhagic telangiectasia of the posterior ⅓ tongue. Telangiectasia (1) are within the mucous membrane of the posterior one-third of the tongue (2), above the epiglottis (3) (see figures **104, 105, 127, 195**).

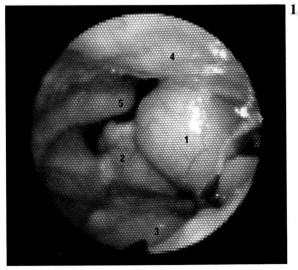

160 Vallecular cyst. A large vallecular cyst (1) lies between the epiglottis (2) and the base of the tongue (3), and is touching the posterior wall of the oropharynx (4). The right ary-epiglottic fold (5) is clearly visible, but the left ary-epiglottic fold is obscured by the cyst.

161

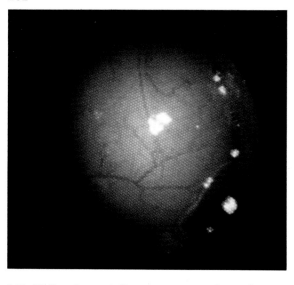

161 Vallecular cyst. Prominent mucosal vessels overlie the cyst. Vallecular cysts are usually asymptomatic until they attain a large size, and may then cause symptoms referable to the throat (see figure **160**).

162

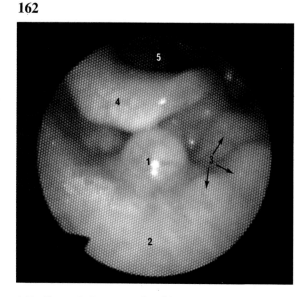

162 Cyst of the posterior ⅓ of tongue. An asymptomatic cyst (1) of the posterior one-third of the tongue (2) is surrounded by healthy lymphoid tissue (3). Beyond the epiglottis (4) is the laryngeal inlet (5).

163

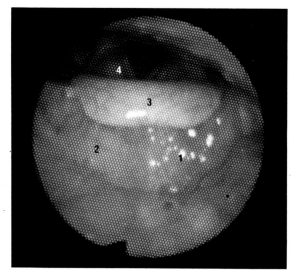

163 Tonsil cyst. A prominent cyst (1), arising from the lower pole of the left tonsil, extends towards the epiglottis (2), and partially fills the vallecula.

164

164 Oedema of the left vallecula. Inflammatory oedema (1) of the left vallecula has resulted from the impaction of a foreign body. The right vallecula (2) is healthy. Beyond the epiglottis (3) are healthy abducted vocal cords (4).

165

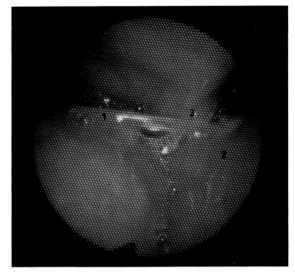

165 Foreign body in the base of tongue. A fish bone (1) is impacted in the base of the tongue (2) and overlies the median glosso-epiglottic fold (3).

166

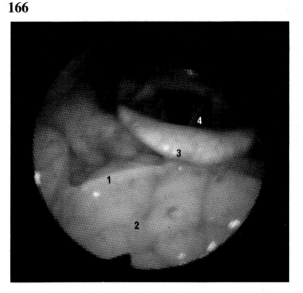

166 Foreign body in the base of tongue. A fish bone (1) is impacted in the base of the tongue (2). The bone lies superior to the epiglottic tip (3), beyond which are abducted vocal cords (4).

167

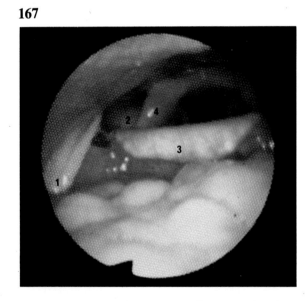

167 Foreign body in the base of tongue. A celery segment (1) is impacted in the right side of the base of the tongue and extends inferiorly into the right pyriform fossa (2). The foreign body is lateral to the epiglottis (3), and the ary-epiglottic fold (4).

168

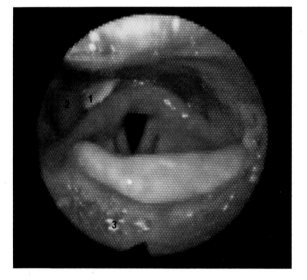

168 Foreign body in the right pyriform fossa. A lamb bone (1) is impacted in the right pyriform fossa (2), making swallowing painful and difficult, and has resulted in pooling of saliva in the valleculae (3).

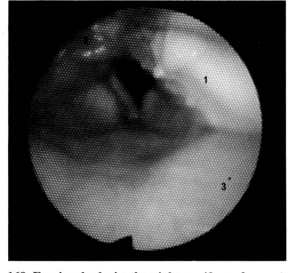

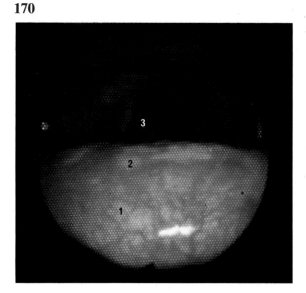

169 Foreign body in the right pyriform fossa. A 5 cm chicken bone (1) is impacted in the right pyriform fossa (2) and is extending upwards, towards the left side of the oropharynx. The bone is in contact with the left side of the epiglottis (3).

170 Epiglottis-swallowed corrosive. Following accidental ingestion of a caustic corrosive, the epiglottis has become acutely inflamed (1), with an overlying exudate (2). The larynx (3) distally is also affected (see figures **145, 207**).

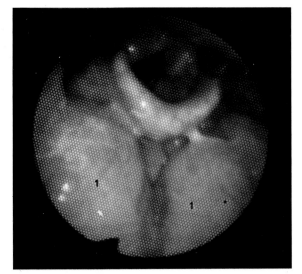

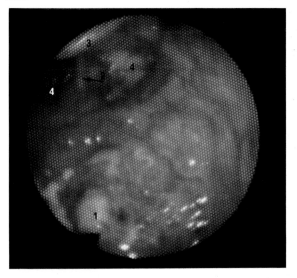

171 Non-Hodgkin's lymphoma of the tongue. The lymphoid tissue of the posterior one-third of the tongue is diffusely enlarged (1) owing to a non-Hodgkin's lymphoma. This lingual tissue may enlarge to such a degree as to cause respiratory obstruction necessitating a tracheostomy.

172 Carcinoma of the tongue. A nodular neoplasm occupies the posterior one-third of the left side of the tongue (1). The median glosso-epiglottic fold (2) connects the tongue to the epiglottis (3) and forms the medial border of both valleculae (4).

173

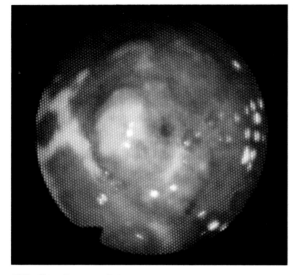

173 Carcinoma of the tongue. A closer view demonstrates the neoplasm to be ulcerating. Histological examination showed this lesion to be a squamous cell carcinoma (see figure **172**).

174

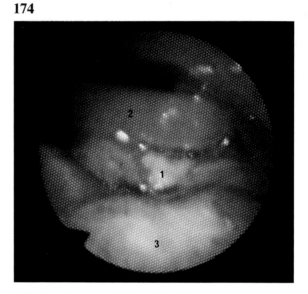

174 Carcinoma of the vallecula. A squamous cell carcinoma (1) lies between an oedematous epiglottis (2), and the posterior one-third of the tongue (3).

175

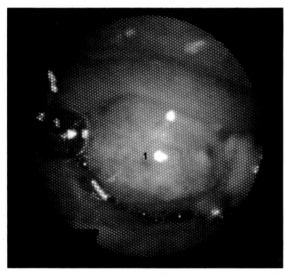

175 Carcinoma of the vallecula. The neoplasm (1) appeared more prominent during protrusion of the tongue (see figure **174**).

176

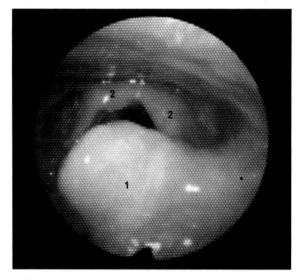

176 Oedema of the epiglottis. A grossly oedematous and asymmetrical epiglottis (1) has resulted from recent radiotherapy to the larynx. This radiotherapy reaction has affected the arytenoid cartilages (2) to a lesser degree.

177

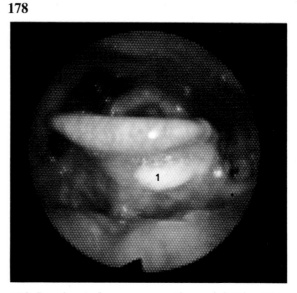

177 Oedema of the epiglottis. Following recent radiotherapy to the larynx, the right half of the epiglottis (1) has become oedematous, while the left half (2) has been spared. An associated reaction has caused gross oedema of the arytenoid cartilages (3). There is no reaction on the posterior wall of the laryngopharynx (4) (see figure **247**).

178

178 Inspissated mucus on the epiglottis. This 'slough' (1), overlying the lingual surface of the epiglottis, was inspissated mucus.

The patient had received radiotherapy to this area 10 years previously.

179

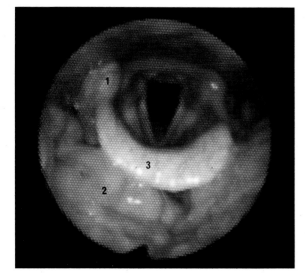

179 Cyst of the right ary-epiglottic fold. The laryngeal inlet is asymmetrical owing to a benign cyst (1) within the free margin of the ary-epiglottic fold. A prominent lingual tonsil (2) is touching the epiglottis (3).

180

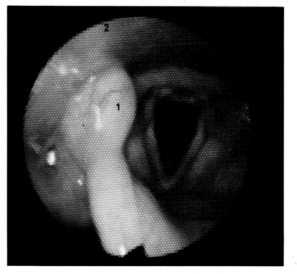

180 Cyst of the right ary-epiglottic fold. Prominent vessels overlie the distended mucous membrane of the cyst (1), which touches the posterior wall of the oropharynx (2) (see figure **179**).

Larynx and laryngopharynx

181

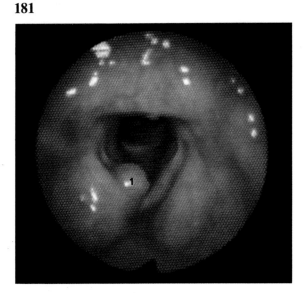

182

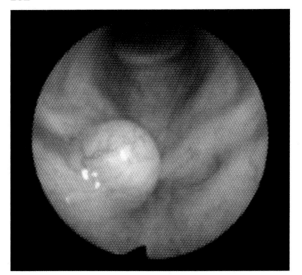

181 Cyst of the right laryngeal sinus. A pale cystic swelling (1) arises from the anterior end of the right laryngeal sinus.

182 Cyst of the right laryngeal sinus. The vessels overlying the cyst are prominent (see figure **181**).

183

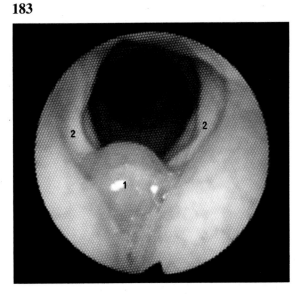

184

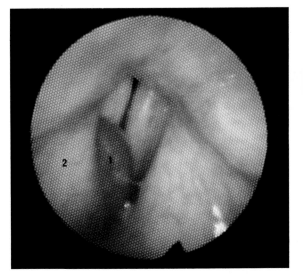

183 Cyst of the right laryngeal sinus during inspiration. A blue cyst (1) arises at the anterior end of the right laryngeal sinus, and is seen during deep inspiration with the vocal cords (2) fully abducted.

184 Cyst of the right laryngeal sinus during phonation. When asked to phonate, there is full adduction of the vocal cords beneath the cyst (1). The cyst underlies the right ventricular band (2) (see figure **183**).

185

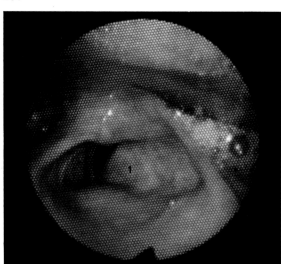

185 Cyst of the left laryngeal sinus. An overall view of the larynx shows a cyst (1) lying above the vocal cords.

186

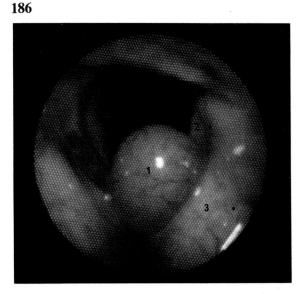

186 Cyst of the left laryngeal sinus. A cyst (1) is arising from the left laryngeal sinus, above the vocal cord (2), and beneath the ventricular band (3).

Supraglottic cysts usually present with hoarseness, but may be asymptomatic (see figure **185**).

187

187 Cyst of the left laryngeal sinus. A large cystic swelling (1) arises in the left laryngeal sinus and is displacing the left ventricular band, and the left vocal cord is obscured. The right vocal cord (2) is healthy.

188

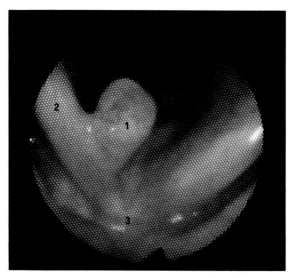

188 Cyst of the right vocal cord during inspiration. A pedunculated cyst (1) is arising from the right vocal cord (2), close to the anterior commissure (3).

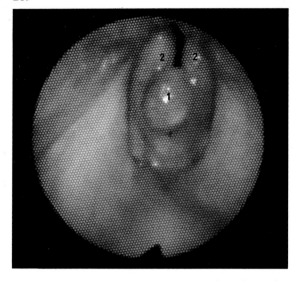

189 Cyst of the right vocal cord during phonation.
During phonation, a pedunculated cyst (1) prolapses superiorly above adducted vocal cords (2) (see figure **188**).

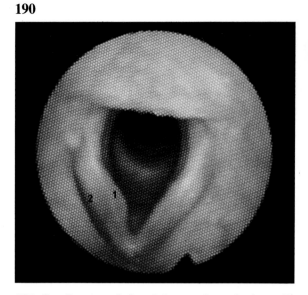

190 Sessile cyst of the right vocal cord. A sessile cyst (1) occupies most of the length of the right vocal cord (2).

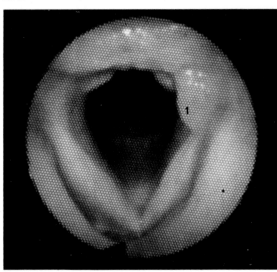

191 Cyst of the left vocal process. This uncommon cyst (1) occupies the medial and superior aspect of the mucous membrane overlying the left vocal process of the arytenoid cartilage.

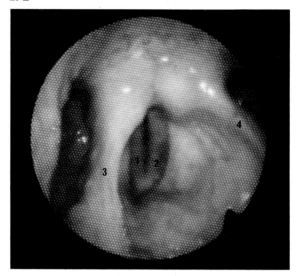

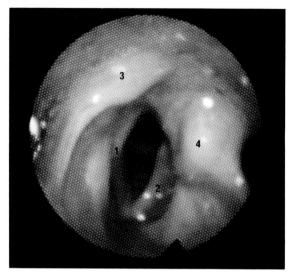

192 Left vocal cord paralysis during phonation.
During phonation, the normal right vocal cord (1) adducts to the midline while the left vocal cord (2) appears bowed and lies at a lower level than the right vocal cord. Adduction to the midline is partial resulting in incomplete glottic closure. Hence the patient will have a breathy voice and a bovine cough.

The right ary-epiglottic fold (3) is tense and appears in the normal position during phonation, whereas the left ary-epiglottic fold (4) has lost its muscle tone and cannot therefore conform to the normal position for phonation.

193 Left vocal cord paralysis during inspiration.
During inspiration, the normal right vocal cord (1) abducts from the midline in the normal manner, while the left vocal cord (2) adopts the same bowed appearance as in phonation. The right arytenoid cartilage (3) is abducted by its lateral crico-arytenoid muscle, whereas on the left side this muscle is paralysed and the left arytenoid cartilage (4) remains stationary (see figure **192**).

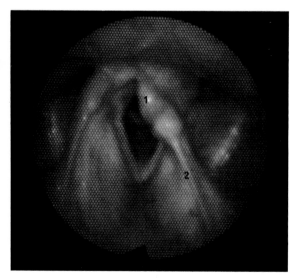

194 Left vocal cord paralysis during inspiration. A left recurrent laryngeal nerve palsy has resulted in paralysis of the laryngeal muscles, allowing the left arytenoid cartilage (1) and left ary-epiglottic fold (2) to prolapse towards the midline.

Paralysis of the left vocal cord is four times more common than paralysis of the right vocal cord because the left recurrent laryngeal nerve has a longer course, passing below the ligamentum arteriosum in the chest, while the right recurrent laryngeal nerve passes below the subclavian artery in the neck.

The intrinsic laryngeal muscles are supplied by the recurrent laryngeal nerve except for the crico-thyroid muscle which is supplied by the external branch of the superior laryngeal nerve.

195

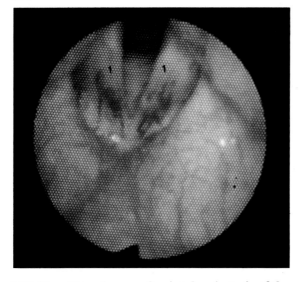

196

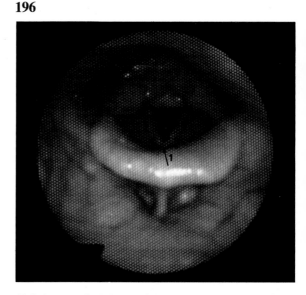

195 Hereditary haemorrhagic telangiectasia of the vocal cords. This persistent appearance of dilated vessels on the superior surface of the anterior half of both vocal cords (1), is due to hereditary haemorrhagic telangiectasia (Osler's disease) (see figures **104, 105, 127, 159**).

196 Acute viral laryngitis. An overall view of the larynx shows generalised acute inflammation of the mucous membrane. Both vocal cords are hyperaemic with an inflammatory exudate at the anterior commissure (1). The supraglottis is also involved in the inflammatory reaction. This response has occurred as part of a generalised acute respiratory tract infection.

197

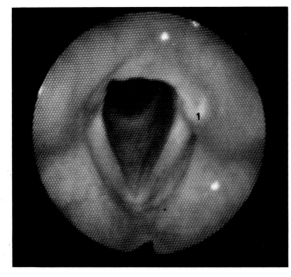

198

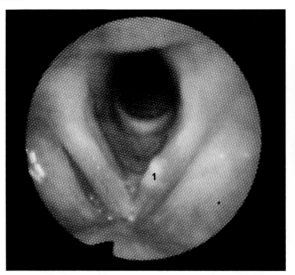

197 Solitary aphthous ulceration of the larynx. Following an upper respiratory tract infection, this patient presented with a painful throat and hoarseness. The left arytenoid cartilage shows a punched-out ulcer (1) on its medial aspect. The ulcer had completely resolved 6 weeks later. The vocal cords are normal.

198 Solitary aphthous ulceration of the larynx. This patient had persisting hoarseness following influenza. A punched-out ulcer (1) lies midway along the left vocal cord. The ulcer had resolved 1 month later.

Laryngeal aphthous ulceration may occur in patients with a history of recurring oral aphthous ulceration.

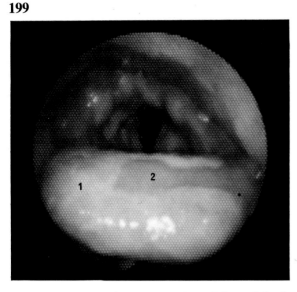

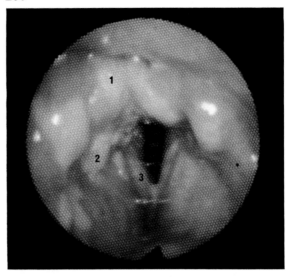

199 Multiple aphthous ulceration of the epiglottis and larynx. Confluent areas of ulceration (1) overlie the laryngeal surface of the epiglottis (2). Distally, the larynx is also affected.

200 Multiple aphthous ulceration of the larynx. Confluent areas of aphthous ulceration have affected the mucous membrane overlying the arytenoid cartilages (1), and the right ventricular band (2). The vocal cords (3) are erythematous, and strands of mucus are bridging the glottis.

The patient complained of severe sore throat and dysphagia following a recent upper respiratory tract infection.

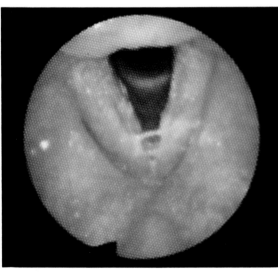

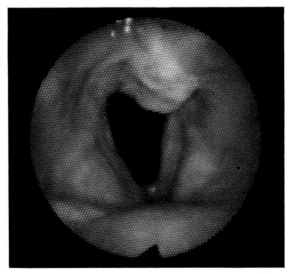

201 Acute bacterial laryngitis. The vocal cords are thickened, red, and oedematous, with an overlying exudate, in response to an acute bacterial infection of the respiratory tract.

202 Acute bacterial laryngitis. Both vocal cords are reddened in response to acute bronchitis, with excessive mucus production. A plug of mucopus has failed to clear the larynx completely during expectoration and has lodged in the inter-arytenoid region, causing secondary bacterial laryngitis.

203

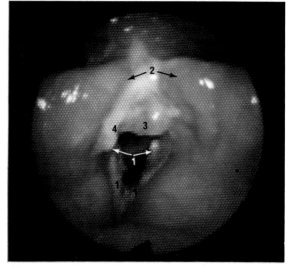

203 Acute bacterial laryngitis. Crusts overlie acutely inflamed vocal cords (1). The mucous membrane overlying the arytenoid cartilages (2) and posterior commissure (3) is also involved in the acute inflammatory reaction. Free-flowing pus (4) courses along the medial border of the right arytenoid cartilage.

204

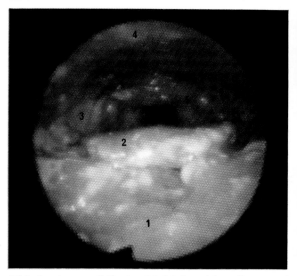

204 Acute candidiasis. Multiple white plaques of *Candida albicans* overlie the inflamed mucous membrane of the tongue (1), epiglottis (2), pyriform fossae (3), posterior pharyngeal wall (4), and supraglottis (5).

205

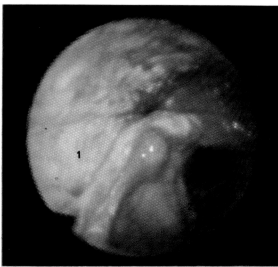

205 Acute candidiasis. This fungal infection is especially pronounced at the right pyriform fossa (1). The vocal cords (2) have been spared. Candidiasis may occur in patients taking broad-spectrum antibiotics or steroids, in diabetics, or in other debilitating states (see figure **204**).

206

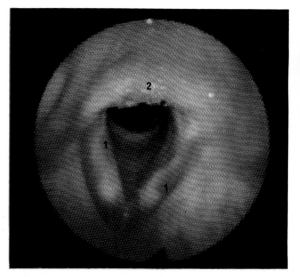

206 Acute chemical laryngitis. Following long-term exposure to toxic industrial fumes, this patient presented with hoarseness, and examination showed patchy submucosal haemorrhages (1) along both vocal cords.

The inflammatory reaction has also affected the supraglottic mucous membrane, and the inter-arytenoid area (2).

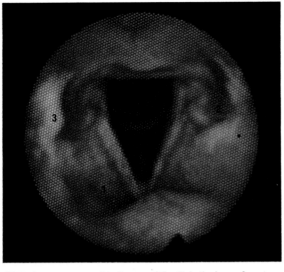

207 Acute corrosive laryngitis. Inhalation of a caustic corrosive has resulted in areas of superficial ulceration (1), haemorrhage (2), and exudate (3) (see figures **145, 170**).

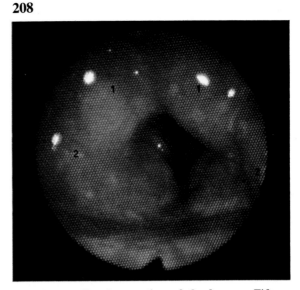

208 Acute allergic reaction of the larynx. Fifteen minutes following a meal of shellfish, this previously well patient suddenly developed acute inspiratory stridor. Examination showed gross oedema of the mucous membrane overlying the arytenoid cartilages (1) and the ary-epiglottic folds (2). The vocal cords were not involved in the allergic reaction.

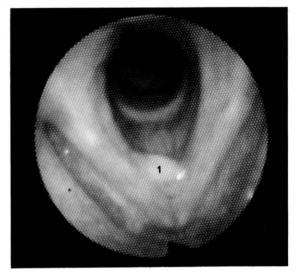

209 Acute allergic reaction of the larynx. Twelve hours later, the allergic reaction had almost resolved following treatment with parenteral hydrocortisone and antihistamines. Submucosal haemorrhages remained throughout the supraglottis (see figure **208**).

210 Mild chronic laryngitis. The vocal cords are thickened, and a globule of sticky mucus (1) lies at the anterior commissure.

This appearance may occur in patients with chronic bronchitis.

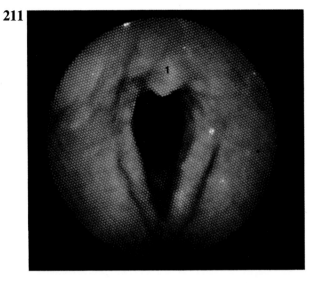

211 Generalised chronic laryngitis. The entire laryngeal mucous membrane is reddened and thickened, particularly at the posterior commissure (1). This appearance has occurred as a consequence of chronic irritation.

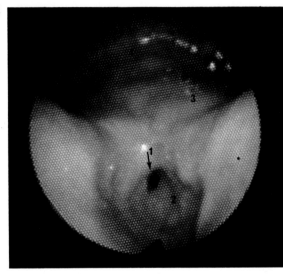

212 Laryngeal scleroma. This appearance represents the healing end stage of laryngeal scleroma, and has resulted in almost complete obliteration of the laryngeal airway. The residual air inlet (1) is at the level of the ventricular bands (2).

The vocal cords cannot be seen and lie below the level of this aperture, at the level of the arytenoid cartilages (3), which lie anterior to the post-cricoid space (4). This patient required a permanent tracheostomy.

Scleroma is caused by the bacterium *Klebsiella scleromatis*, and occurs more commonly in Eastern Europe, especially Poland, and in South America, especially Venezuela.

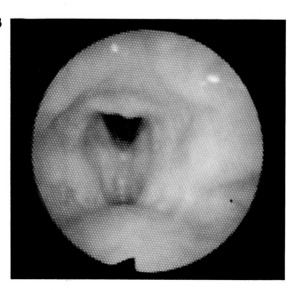

213 Healed tuberculosis of the larynx. A web occludes the anterior two-thirds of the glottis, and is a sequel of successfully treated tuberculosis as a child, many years previously.

The airway was adequate, and a tracheostomy was not required.

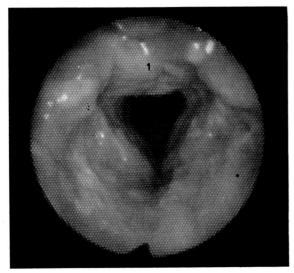

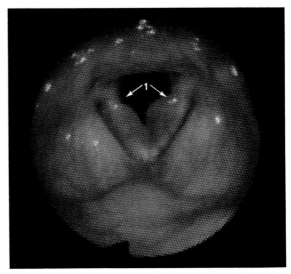

214 Laryngeal sarcoidosis. The entire laryngeal supraglottic mucous membrane is reddened, thickened and granular, and is a localised manifestation of this systemic disease. Raised mucosal ridges are pronounced at the posterior commissure (1).

215 Reinke's Oedema. Gross oedema within the subepithelial space of both vocal cords along their entire lengths (1) has produced this extreme polypoidal appearance.

This may be a presenting feature of hypothyroidism.

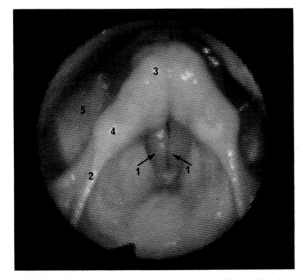

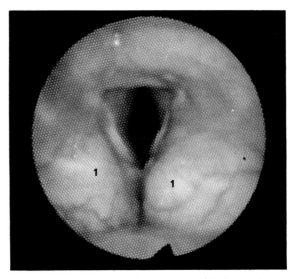

216 Reinke's oedema during phonation. During phonation, the oedematous tissue (1) appears above both vocal cords, which are adducted. This extensive polypoidal tissue vibrates above the glottis during expiration, and produces a characteristically coarse voice during phonation.

Anatomical features particularly well shown are the ary-epiglottic folds (2), the corniculate (3) and cuneiform (4) accessory laryngeal cartilages, the pyriform fossae (5), and the post-cricoid space (6) (see figure **215**).

217 Hypertrophied ventricular bands. In dysphonia plicae ventricularis, the ventricular bands are pathologically hypertrophied (1). The vocal cords are normal in structure and function.

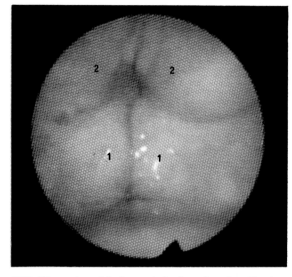

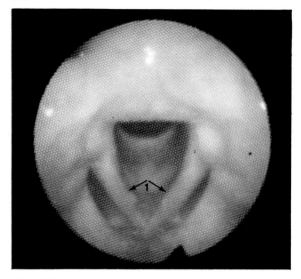

218 Hypertrophied ventricular bands. During phonation an abnormal voice is produced by the apposition of the ventricular bands (1) and arytenoid cartilages (2) above the vocal cords. Consequently, the vocal cords cannot be visualised at laryngoscopy during phonation.

Although this is an abnormal way of producing a voice, this is a part of the physiological sphincter mechanism which protects the lower respiratory tract.

219 Bilateral vocal cord nodules (singers' nodules). These benign nodules (1) characteristically occur at the junction of the anterior one-third and the posterior two-thirds of the glottic aperture (which is midway along the true vocal cord).

The nodules are classically situated at the point of maximum impact of the adducting vocal cords, and usually occur as a result of vocal abuse.

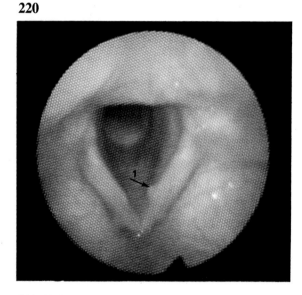

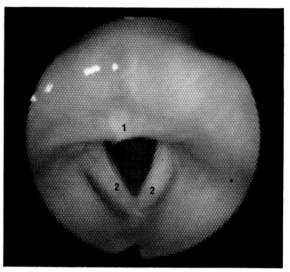

220 Unilateral vocal cord nodule (singers' nodule). Vocal cord nodules may be bilateral or unilateral (1).

Speech therapy treatment is usually successful in eliminating these nodules, but they may be removed surgically in the more resistant case.

221 Synechia of the posterior commissure. A dense fibrous web (1) at the posterior commissure is the end result of burns, due to the inhalation of hot smoke as a child when trapped in a house fire. The vocal cords (2) have been spared.

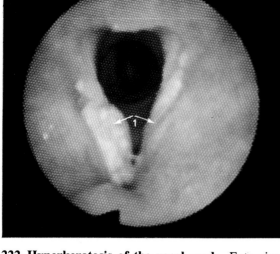

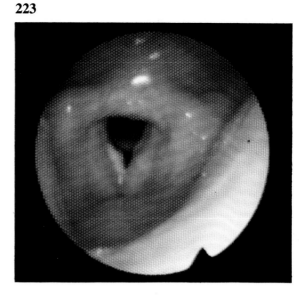

222 Hyperkeratosis of the vocal cords. Extensive white hyperkeratotic squamous epithelium overlies both vocal cords (1). The bifurcation of the trachea into the left and right main bronchi appears distally.

This form of laryngitis usually occurs following chronic irritation, and is a pre-malignant condition.

223 Leukoplakia of the right vocal cord. A white patch overlies the right vocal cord. This may conceal an underlying neoplasm, and regular follow-up will be required.

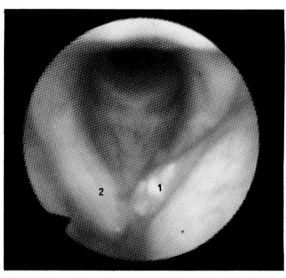

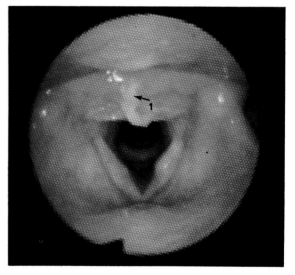

224 Leukoplakia of the left vocal cord. A localised white patch is at the anterior end of the left vocal cord (1). The right vocal cord (2) is healthy.

This appearance may overlie a squamous cell carcinoma.

225 Inter-arytenoid pachydermia. Pachydermia, or inter-arytenoid hyperkeratosis, is present in a vertical strip at the posterior commissure (1).

226

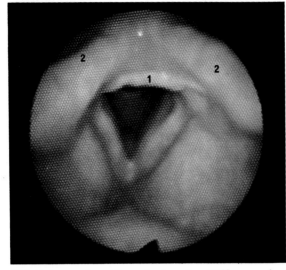

226 Inter-arytenoid pachydermia. Diffuse pachydermia (1) extends between the arytenoid cartilages (2).

227

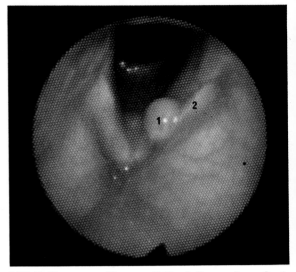

227 Solitary papilloma of the left vocal cord. A solitary papilloma (1) arises midway along the left vocal cord (2). Such solitary lesions are not of viral aetiology.

228

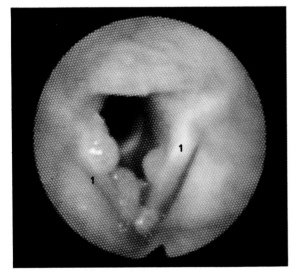

228 Multiple papillomatosis (juvenile papillomatosis) during inspiration. Multiple sessile papillomata arise along both vocal cords (1), in an adolescent. This condition is thought to be of viral aetiology, and the papillomata may appear throughout the respiratory tract. Spontaneous regression can occur.

229

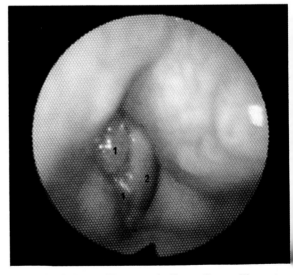

229 Multiple papillomatosis (juvenile papillomatosis) during phonation. During phonation, the papillomata (1) prolapse superiorly. The papillomata are less extensive on the left vocal cord (2). Hoarseness is usually severe (see figure **228**).

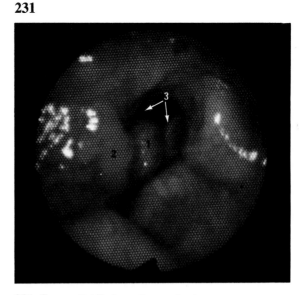

230 Supraglottic plasmacytoma. A bulky neoplasm (1) is arising from the left ventricular band (2), and obscures the left vocal cord, while the right vocal cord (3) is healthy.

This is a solitary lesion but may occasionally be associated with systemic disease (multiple myeloma).

231 Supraglottic lymphoma during inspiration. A non-Hodgkin's lymphoma (1) arises in the right laryngeal sinus and is extending into the right ventricular band (2). The vocal cords (3) remain free of tumour.

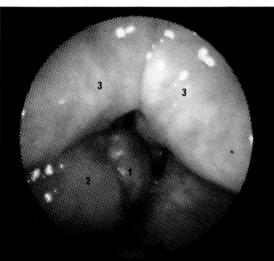

232 Supraglottic lymphoma during phonation. During phonation, the glottic aperture is partially obscured by the neoplasm (1). The right ventricular band (2) and arytenoid cartilages (3) are oedematous.

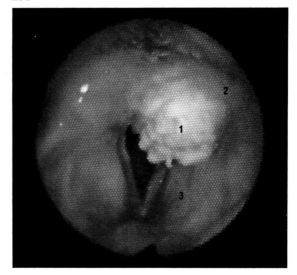

233 Supraglottic squamous cell carcinoma. A proliferative and keratinising squamous cell carcinoma (1) overlies the left arytenoid cartilage (2) and the left ventricular band (3).

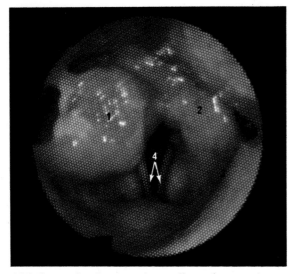

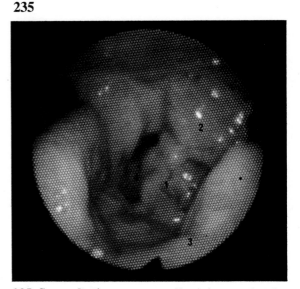

234 Supraglottic squamous cell carcinoma. A proliferative and ulcerating neoplasm (1) is arising on the right arytenoid cartilage, and has invaded the right ary-epiglottic fold.

The left arytenoid cartilage (2), the left ary-epiglottic fold (3), and the vocal cords (4) are not involved.

235 Supraglottic squamous cell carcinoma. A nodular neoplasm arises from the left ventricular band (1). The left arytenoid cartilage (2) and epiglottis (3) are oedematous as a result of sub-mucosal extension of the neoplasm. The glottic aperture (4) is partially obscured.

The left hemilarynx was fixed and resulted in stridor.

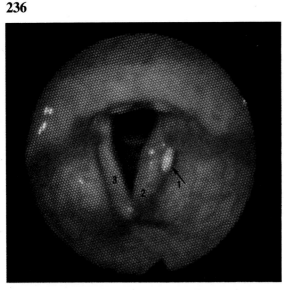

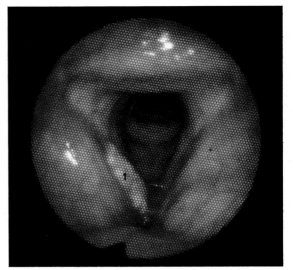

236 Laryngeal sinus squamous cell carcinoma. A squamous cell carcinoma in the floor of the laryngeal sinus (1) resulted in a unilateral red left vocal cord (2), which is infiltrated with tumour. The right vocal cord appears normal (3). The mobility of the left vocal cord was reduced.

237 Glottic squamous cell carcinoma. The right vocal cord is infiltrated along its length by a keratinising squamous cell carcinoma (1). Both vocal cords are fully mobile.

238

239

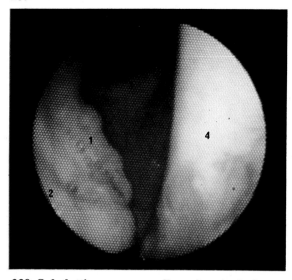

238 Glottic squamous cell carcinoma. The right vocal cord is infiltrated along its length by a nodular non-keratinising squamous cell carcinoma (1), which extends onto the vocal process of the arytenoid cartilage (2), and into the laryngeal sinus. The right vocal cord, however, remains fully mobile.

239 Subglottic squamous cell carcinoma. An irregular and ulcerated lesion (1) lies beneath the normal right vocal cord (2), and extends towards the cricoid cartilage (3). The left vocal cord (4) is unaffected.

240

241

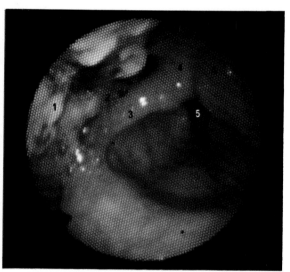

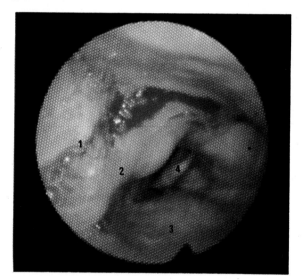

240 Lateral pharyngeal wall squamous cell carcinoma. An extensive, irregular and ulcerated neoplasm (1) covers the right lateral wall of the pharynx and extends into the right pyriform fossa (2). The mucous membrane overlying the right ary-epiglottic fold (3) and the arytenoid cartilages (4) is oedematous. The vocal cords (5) are adducted.

241 Pyriform fossa squamous cell carcinoma. A friable and ulcerating lesion (1) fills the right pyriform fossa and involves the right ary-epiglottic fold (2), which is displaced towards the midline. The epiglottic tubercle (3) and the vocal cords (4) are unaffected.

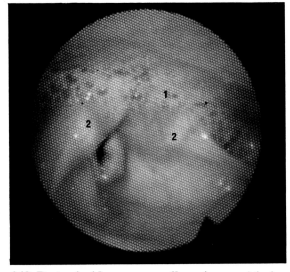

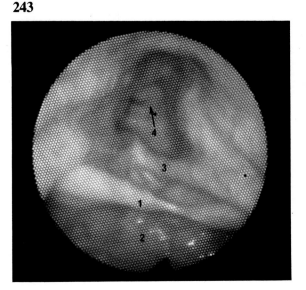

242 Post-cricoid squamous cell carcinoma. A lesion limited to the post-cricoid area, which cannot be visualised with the FFRL, has produced this typical sign of pooling of saliva (1) in the laryngopharynx. The saliva is encroaching upon the arytenoid cartilages (2). Pooling of saliva in the laryngopharynx may also occur when foreign bodies are lodged in this area, or in neuromuscular disorders affecting the swallowing mechanism.

243 The pharynx following laryngectomy. A horizontal bar of pharyngeal mucosa (1) is a common appearance following a vertical repair of the pharyngeal defect after the operation of total laryngectomy. Above this bar is the healthy mucosa of the tongue (2), while below, the mucous membrane of the laryngopharynx (3) funnels towards the oesophagus (4).

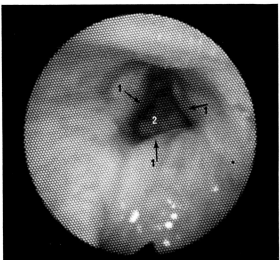

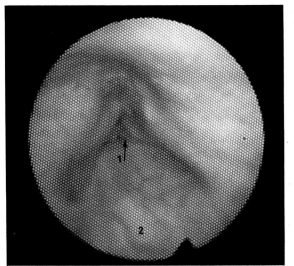

244 The pharynx following laryngectomy. When asked to swallow, the sphincter at the upper end of the oesophagus initially relaxes (1), allowing visualisation of the oesophagus (2) (see figure **243**).

245 The pharynx following laryngectomy. At rest the upper oesophageal sphincter is contracted and the oesophagus is closed. A pin-point aperture is all that remains (1). A healthy suture line (2) of the pharyngeal repair lies anteriorly (see figure **243**).

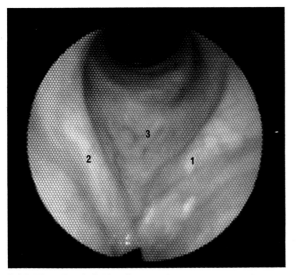

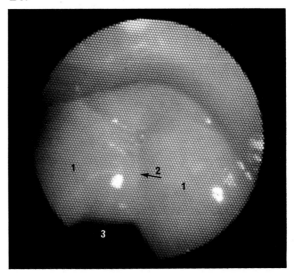

246 The glottis following cordectomy. The left vocal cord was excised for a verrucous carcinoma. Subsequent scarring has resulted in the formation of a new vocal cord (1). The right vocal cord (2) and subglottis (3) are normal.

247 Oedema of the arytenoid cartilages. The mucous membrane overlying the arytenoid cartilages is lax and this may result in gross oedema to this area (1) following radiotherapy.

The posterior commissure (2) is a groove, and anterior to this is the glottis (3) (see figure **177**).

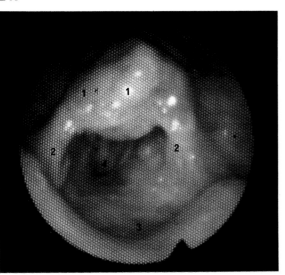

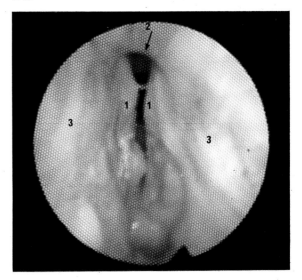

248 Mucositis of the larynx. This early change following radiotherapy shows an inflammatory response of the mucous membrane over the arytenoid cartilages (1) and ary-epiglottic folds (2) as far as the epiglottis (3). The vocal cords (4) are minimally affected by this reaction.

249 Laryngitis sicca. Both vocal cords (1) are covered by a dry and atrophic mucous membrane with overlying crusts, and are fixed in adduction. A tracheostomy was required. A strand of mucus lies towards the posterior commissure (2). The mucous membrane of the ventricular bands (3) is pale and atrophic.

250

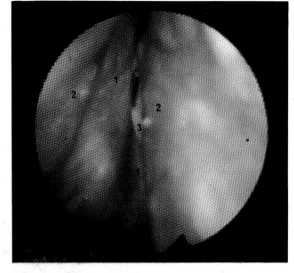

250 Laryngitis sicca. Crusts overlie the vocal cords (1) and the ventricular bands (2). An area of hyperkeratosis (3) overlies the mid-section of the left vocal cord. Laryngitis sicca may be a late complication of radiotherapy.

251

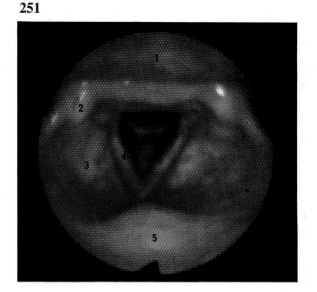

251 Laryngeal telangiectasia. Telangiectasia, as a late consequence of radiotherapy, may appear throughout the larynx and are present on the posterior wall of the laryngopharynx (1), the arytenoid cartilages (2), the ventricular bands (3), the vocal cords (4), and on the laryngeal surface of the epiglottis (5).

252

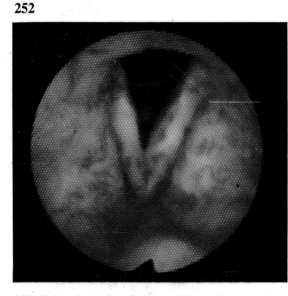

252 Telangiectasia of the vocal cords. Prominent telangiectasia overlie the vocal cords and the ventricular bands (see figure **251**).

253

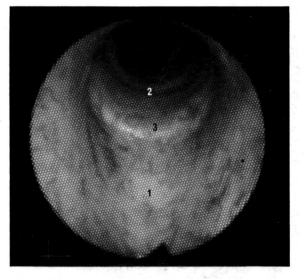

253 Telangiectasia of the subglottis and trachea. The mucous membrane of the subglottis (1) and trachea (2) shows extensive telangiectasia following radiotherapy.

The anterior arch of the cricoid cartilage (3) is normally prominent (see figure **251**).

254

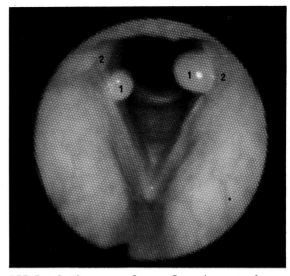

255

254 Synechia of the posterior commissure. A fibrotic inter-arytenoid web (1) has occurred as a late sequel of radiotherapy.

The vocal cords (2) and the ventricular bands (3), are covered by pale and atrophic mucous membrane containing telangiectasia.

255 Intubation granuloma. Opposing granulomas (1) lie over the vocal processes of both arytenoid cartilages (2). This patient was rendered unconscious following a road traffic accident, and was intubated as a matter of urgency upon arrival at hospital. He was then ventilated for a week. This was the appearance 6 weeks later.

256

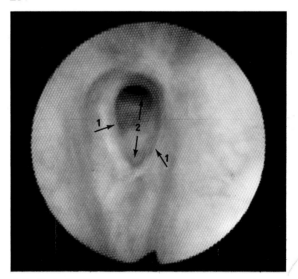

257

256 Subglottic stenosis. A benign fibrous stricture (1) is in the subglottic space immediately below healthy vocal cords (2).

257 Subglottic stenosis. Severe fibrous scarring (1) has resulted in stricture formation in the subglottic space. The stricture extends over a length of 1 cm (‹--2--›) (see figure **256**).

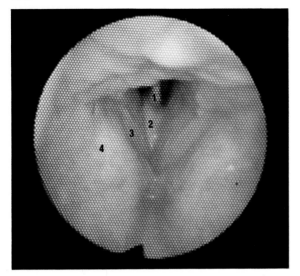

258 Foreign body in the subglottis. An inhaled lamb bone (1) has lodged anteriorly in the subglottic space (2), below the vocal cords (3) and the ventricular bands (4). There is considerable subglottic oedema.

Trachea

259

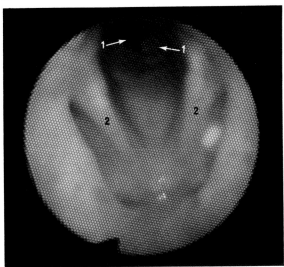

260

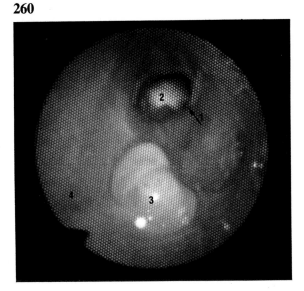

259 Tracheal stenosis. A post-intubation stricture (1) lies in the upper trachea, and is distal to healthy vocal cords (2).

260 Tracheal stenosis. Beyond the stricture (1) is a white tracheostomy tube (2). Polypoidal granulations (3) are above this stricture, and arise from inflamed mucous membrane (4) (see figure **259**).

261

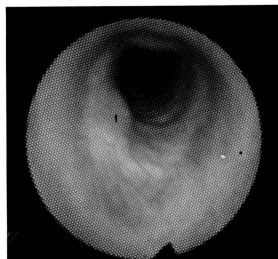

262

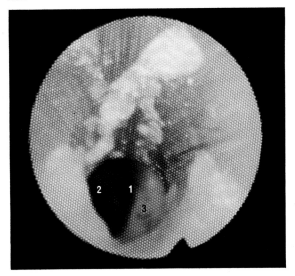

261 Tracheal chondroma. A benign lesion (1) lies anteriorly on the medial aspect of the right side of the cricoid cartilage.

262 Silver tracheostomy tube. On the inner surface of a silver tracheostomy tube lies dried mucus. The origin of the left (1) and right (2) main bronchi lies at the level of the carina, beyond the trachea (3).

References

1. Lancer J.M., Moir A.A. (1985) The flexible fibreoptic rhinolaryngoscope. *Journal of Laryngology and Otology*; **99**:767–770.

2. Lancer J.M., Jones A.S. (1985) Flexible fibreoptic rhinolaryngoscopy: Results of 338 consecutive examinations. *Journal of Laryngology and Otology;* **99**: 771-773.

3. Lancer J.M. (1986) Photography and the flexible fibreoptic rhinolaryngoscope. *Journal of Laryngology and Otology*; **100:** 41-46.

Index

N.B. Figures refer to picture numbers.